Forestry Commission Booklet No. 45

D1784023

Standard Time Tables and Output Guides

London
Her Majesty's Stationery Office

Acknowledgements

The Forestry Commission is indebted to past and present members of its Work Study Branch for the preparation of this Booklet. W. O. Wittering, Chief Work Study Officer, helped considerably with the compilation. The forest staff and their Trades Unions have given very valuable help and co-operation with the basic studies, their development and application.

ISBN 011 710037 4

Preface

The demand from private forestry for guidance on payment of workers has increased enormously over the past few years and the Forestry Commission has been meeting this demand by giving enquirers copies of its Standard Time Tables and Output Guides. The indiscriminate issue of these tables is however likely to cause more problems than it solves, particularly where the user has little experience of using such tables as a basis for incentive payment, or, what is worse, he uses outdated tables.

To meet these needs, it has been decided to publish current tables in the form of a Forestry Commission Booklet every few years. This issue is dated 1978; users are strongly recommended in their own interests to replace it by the next edition as soon as it is available.

It will be appreciated that some knowledge of work study and particularly its application to the field of incentive payment is of value in using these tables. Some guidance is given in this Booklet but readers who feel they need more guidance are advised to read Forestry Commission Bulletin No. 47 *Work Study in Forestry*. For those who need more personal advice, they can contact the Forestry Commission Work Study Headquarters or the nearest Work Study Team Leader whose addresses are given on p xxxi. It is regretted that only limited advice can be given; the staff are not normally available to undertake contract work study.

The Forestry Commission is publishing these tables as a service to private forestry. It disclaims responsibility for any inaccuracies in them and for any losses which may arise from their use.

Alice Holt, Hants.
December 1978

List of Contents

What is Standard Time?

The total time required to complete a job is a combination of:
  (i)   direct work
  (ii) indirect work
  (iii) rest

Taking these in turn:

  (i)   **Direct work** (also known as 'Cyclic Work') is the specific job to be done, for example, in hand weeding it is the actual cutting of the weeds; in thinning it is felling and delimbing a tree. During the preparation of Standard Times, this work is timed in detail and relationships between the time taken and those factors which determine the work content of the job (the volume of the tree, weed type, etc) are established.

  (ii) **Indirect work** (also known as 'Other Work') consists of those things which are not done directly to the trees (or, in weeding, to the weeds) but which have to be carried out all the same; such things as sharpening hooks, refuelling machines, etc. These will also have been timed during studies. Indirect work is usually expressed as a percentage of direct work.

  (iii) **Rest** is necessary in all forms of work, particularly the heavy manual work found in forestry. The heavier the work, the more rest is required. The amount of rest which should be taken in each job is calculated from tables based on experience in a variety of industries including forestry, and checked against the rest actually taken by men during time studies. Standard Time Tables and Output Guides include rest allowances based on factors such as energy output, posture, motions, noise and vibration, dirt and toxic fumes, etc. Rest is expressed as a percentage of the total work, ie direct work plus indirect work, to be done.

These three factors added together give Standard Time, the time in which a job can be done by an average skilled and properly trained man.

Presentation of Standard Time to Men and Management—Standard Time Tables

All Standard Time Tables have standard headings:

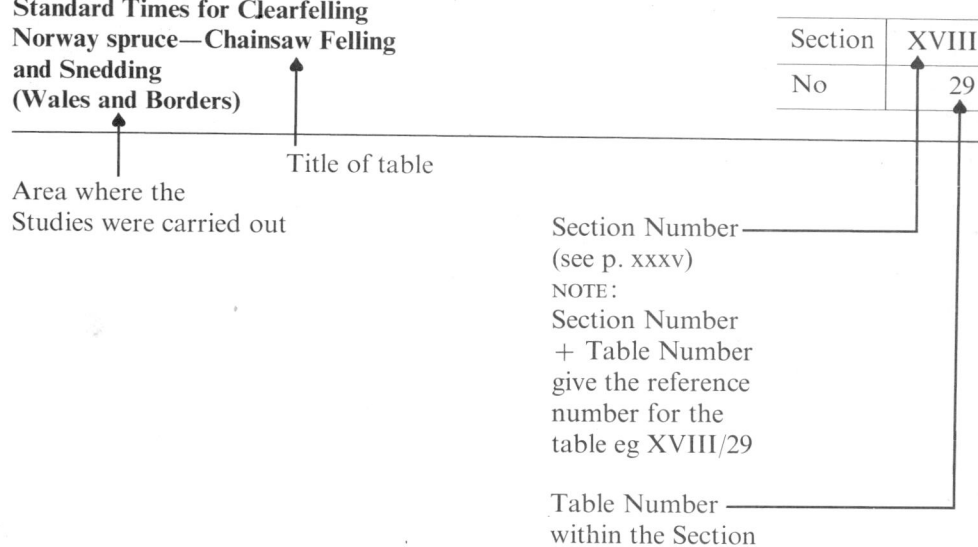

Standard Times for Clearfelling Norway spruce—Chainsaw Felling and Snedding (Wales and Borders)	Section	XVIII
	No	29

Area where the Studies were carried out

Title of table

Section Number—
(see p. xxxv)
NOTE:
Section Number
+ Table Number
give the reference
number for the
table eg XVIII/29

Table Number —
within the Section

The time for the job is presented in the form of a Standard Time Table which, in the Forestry Commission, is always laid out in the same form. It contains seven paragraphs in the following order:
(1) Conditions
(2) Job Specification
(3) Tools and Equipment
(4) Allowances
(5) Method of selecting the Standard Time
(6) Tables of Standard Times
(7) Modifications and variations to the Standard Times.
Each of these is now considered in turn.

1 Conditions

The physical conditions under which a particular job is carried out affect the time taken. In industry, conditions are usually controlled and can be modified to suit the work. In forestry some conditions cannot be altered; for example, the slope of the ground. Others can be affected by earlier decisions by management; as, for example, the amount of brashing will affect the times for thinning.

Hence, in every case a broad classification of conditions to which the standard times are applicable is given and attention is drawn to likely variations in conditions which may cause alterations to the standard times. The unmodified conditions themselves are not necessarily constant for a given operation; for example, thinning times for spruce would assume the presence of some drains while times for pine in east England would not.

If all the conditions described are not present then either some modification may be necessary (as given in the last paragraph of the table) or the times may not apply to that job.

In short, the 'Conditions' paragraph tells the worker what he can expect to find when he goes into the compartment to do the work laid down in the job specification.

For example, for clearfelling Norway spruce (Table XVIII/29): The Standard Times apply to stands clearfelled under the following conditions:

(a) The volume of the average tree is determined by the tariff system. When in stands previously crown thinned it is necessary to sub-divide the dbh distribution (see Forestry Commission Booklet 36 paragraph 26 (ix) page 12); a different average tree size should be used for each sub-division when calculating the Standard Times.

(b) The brashing percentage is at least 90% of the measurable trees (see paragraph 7c).

(c) Floor conditions are average for the species, ie normal amounts of lop and top from thinnings, little ground vegetation, few rocks, normal drainage patterns. Slopes up to 25% for areas from which the felled trees are removed during felling (Table 6B) and up to 35% for other areas (Table 6A).

(d) One man working.

2 Job Specification

Quality standards must be laid down and adhered to. It is just as important to see that excessively high and thus expensive standards are not set as it is to see that lower standards do not creep in unnoticed. The inspection of work is a prime function of junior supervisors and they should receive guidance as to the correct standard required. The job specification frequently becomes a condition for the next operation; for example, 'Poles must be extracted and piled with their butts together' is an aid to subsequent cross-cutting. The balancing of tasks to give the lowest overall costs is an important part of the original method study.

A supervisor may well decide that a worker who has failed to meet the required job specification (quality standard) may not be due to receive the full rate for the job until the required standard has been met.

An example of a Job Specification for clearfelling Norway spruce, from Table XVIII/29, is:

The Standard Times apply to the following work:

(a) Trees to be felled in the direction most suitable for extraction or as directed by the supervisor.

(b) Stumps to be cut as low as possible.

(c) Branches and brashing and pruning knots to be cut off flush with the stem of the tree, snedding with the chainsaw. Smaller trees will be turned from the tip for snedding the undersides. Larger trees (above about $0.25\,m^3$) will normally only be snedded as thoroughly as is possible without turning them (but see paragraph 7g).

N.B.: When bigger trees are crosscut at stump into one or more pieces of timber and a top, it is usual to turn the pieces to sned their undersides after crosscutting. Because it is easier to sned the branches after the pieces have been turned no extra time allowance is normally necessary for the extra turning involved (but see paragraph 7h). The crosscutting should be paid for separately.

(d) Lop and top to be cleared from major racks, roads and main drains as required by the supervisor.

(e) Tops to be cut off at a diameter of 7 cm.

(f) Tops to be cut into lengths not exceeding 1.2 m.

(g) All cuts including the felling cut to be made as squarely across the tree as possible.

(h) Buttresses to be removed with the chainsaw after felling.

(i) Stumps to be treated immediately after felling.

N.B. Treating stumps requires extra Standard Time—see paragraph 7f. In some tables, this paragraph is expanded by an Appendix which gives the method of working. It is hoped that by the time the next edition of this Booklet appears, all tables will contain such an Appendix.

3 Tools and Equipment (including safety equipment)

It is essential that the correct equipment be available including both gear which is required occasionally and those items which are needed regularly. It is also essential that a high standard of tool maintenance is achieved and maintained. This paragraph also specifies the safety equipment and clothing necessary.

For example, for clearfelling Norway spruce (Table XVIII/29):

(a) Lightweight anti-vibration chainsaw of an approved pattern and suitable for chainsaw snedding, together with spares and maintenance tools.

(b) Breaking bar which may have a cant-hook attachment.

(c) 13 cm plastic felling wedges as required.

(d) When larger trees are turned from the butt, or timber pieces are turned after crosscutting, a cant-hook or breaking bar with cant-hook attachment (see (b)).

(e) If trees are crosscut at stump, Spencer or Bushman's logger's tape and pulphook.

(f) Suitable equipment for stump treatment.

(g) The following protective clothing:
 Safety Helmet BS2826, 2095 or 5240
 Eye Protection
 Ear Defenders
 Gloves incorporating ballistic nylon lining on back of left hand
 Nylon leg guard
 Safety boots incorporating inner ballistic nylon guard.

N.B. Safety requirements can change; alterations will be noted in new editions of this Booklet.

4 Allowances

In the standard times, allowances are made for ancillary tasks such as tool maintenance, walking from one site to another, counting and checking work, etc. These together with a small allowance for contingencies are usually expressed as a percentage of the basic time for the particular job. This 'other work' allowance and the rest allowed is stated. A reminder is given to management that an incentive allowance must be included in the Price per Standard

Minute and guidance is given on paying for worker-owned tools.

For example, for clearfelling Norway spruce (Table XVIII/29):

Included in the Standard Times:

(a) For contingencies and work other than that performed on individual trees, eg refuelling and day to day maintenance of the saw, clearing lop and top from rides, etc, 23% of the time spent on clearfelling.

N.B. Mechanical breakdowns to the saw in excess of 15 minutes are not included in the allowance and should be paid for at time rates.

(b) For personal needs and rest, 25% of the total working time.

To be included in the Price per Standard Minute:

(c) The appropriate incentive allowance.

(d) For calculating the payment for worker-owned chainsaws, chainsaw utilisation is 70% of the total working time when the chainsaw is running during stump treatment or when stumps are not treated. When the saw is not running during stump treatment chainsaw utilisation is 70% of the standard time for felling or about 65% of the total working time.

Information on how to calculate the Price per Standard Minute is given on page xxi.

5 Method of Selecting the Standard Time

The user is told how to gain entry to the tables of standard times and what measurements he will need to take beforehand. The skills of the work study man and the statistician come together here in deciding which parameters to use to gain the maximum degree of explanation of variation but at the same time not requiring the manager to carry out measurements which he may not need for other purposes. For example, it is in order to produce a table giving times per tonne for loading where the produce is weighed for subsequent invoicing of the customer but it would be quite wrong to insist on all branches of a tree being measured so that the worker can be paid for snedding as this measurement is of no other value to the supervisor.

For example, for clearfelling Norway spruce (Table XVIII/29):

(a) *Different Working Systems*
Two quite distinct types of harvesting operations are covered by the standard times: other types may be thought of as intermediate.

(i) Felled trees *not* extracted immediately after felling. When felling and extraction are two separate operations done at different times the feller is working for most of the time between felled trees: and these and the lop and top caught between them are a considerable hindrance during snedding. This type of working often occurs on slopes where it is followed by double-drum winch extraction but may occur elsewhere. For this type of working use Table 6A.

(ii) Felled trees extracted immediately after felling. This usually occurs when the feller is working along a face or on more than one short face and extraction is close enough behind him for the ground to be cleared of felled trees before he starts along the face again; but it may also occur with other systems. For methods where the cutter is working for most of the time standing on lop and top but on ground cleared of previously felled trees use Table 6B.

(iii) Intermediate between (i) and (ii). Some felling methods are intermediate between the two main types. For such felling see paragraph 7a.

(b) Trees should be tariffed by the method described in Forestry Commission Booklet 36.

(c) To select the Standard Time from Table 6A or Table 6B select the appropriate average volume or volumes for the stand and read off the Standard Time(s) from columns (ii) and/or (iv).

6 Tables of Standard Times

These tables form the central feature of each scheme but are only valid within the conditions, job specification, working method and method of evaluation described.

For example, for clearfelling Norway spruce (Table XVIII/29):

6A *Standard Times for clearfelling Norway spruce—*
felled trees not extracted immediately after felling

Volume of Average Tree in m³	Standard Time per Tree in Minutes	Volume of Average Tree in m³	Standard Time per Tree in Minutes
0·04	4·17	0·72	17·32
0·06	4·74	0·76	17·77
0·08	5·29	0·80	18·19
0·10	5·83	0·84	18·59

Volume of Average Tree in m³	Standard Time per Tree in Minutes	Volume of Average Tree in m³	Standard Time per Tree in Minutes
0·12	6·36	0·88	18·96
0·14	6·87	0·92	19·31
0·16	7·38	0·96	19·64
0·18	7·87	1·00	19·95
0·20	8·35	1·04	20·23
0·24	9·28	1·08	20·50
0·28	10·16	1·12	20·75
0·32	11·00	1·16	20·99
0·36	11·80	1·20	21·21
0·40	12·56	1·24	21·41
0·44	13·27	1·28	21·61
0·48	13·95	1·32	21·79
0·52	14·60	1·36	21·96
0·56	15·21	1·40	22·13
0·60	15·78	1·44	22·29
0·64	16·33	1·48	22·44
0·68	16·84	1·50	22·52

6B   *Standard Times for clearfelling Norway spruce—*
felled trees extracted immediately after felling

Volume of Average Tree in m³	Standard Time per Tree in Minutes	Volume of Average Tree in m³	Standard Time per Tree in Minutes
0·04	2·69	0·52	11.90
0·06	3·23	0·56	12·37
0·08	3·75	0·60	12·80
0·10	4·26	0·64	13·19
0·12	4·76	0·68	13·55
0·14	5·24	0·72	13·88
0·16	5·70	0·76	14·17
0·18	6·15	0·80	14·44
0·20	6·59	0·84	14·68
0·24	7·43	0·88	14·90
0·28	8·21	0·92	15·10
0·32	8·94	0·96	15·27
0·36	9·62	1·00	15·42
0·40	10·26	1·04	15·56
0·44	10·85	1·08	15·68
0·48	11·40	1·10	15·74

7 Modifications and Variations to the Standard Times

Once the main variables have been found and a table of Standard Times produced, it is generally possible to predict what will happen if certain other subsidiary factors change; for example, if there are brambles in a stand to be thinned or if plants to be weeded are very small. Such differences can be allowed for as fixed or percentage alterations to the Standard Times. The accuracy of the allowances for modifications and variations is sometimes less than that of the Main Standard Time Table and they are only to be applied when the prevailing conditions or job specification differ from those listed in paragraphs 1 and 2 of the table.

For example, for clearfelling Norway spruce (Table XVIII/29):

(a) *Other Types of Felling*

Besides felling on to ground where all the previously felled trees are still lying and felling on to ground from which the felled trees have been extracted, there are also felling methods intermediate between these two extremes. For such areas the times in Table 6A should be reduced by up to 20% in steps of 5% according to the supervisor's estimate of the snedding difficulty. The times for felling on to areas with no previously felled poles on the ground (Table 6B) are approximately 20% below those in Table 6A for trees larger than $0.2 \, m^3$, and more than 20% less for smaller trees.

(b) *Ground Conditions*

Where ground conditions are difficult because of rocks, deep drains or excessive steepness (eg slopes in excess of 35%) add up to 10% to the Standard Times in steps of 5%. *Exceptionally* for extremely difficult conditions, an additional 15% may be justified.

(c) *Brashing Percentage*

If the percentage of trees brashed is less than 90% the brashing of all measurable unbrashed trees should be paid for. The time for chainsaw brashing (or minimal brashing and extra snedding) is 0.75 SMs per tree.

(d) *Unmeasurable Trees*

If unmeasurable trees are felled and cut up into 1.2 m lengths the Standard Time for this operation is 1.00 SMs per stem.

N.B. Extra brashing and the disposal of unmeasurable trees may be paid for separately from the felling of the crop. Alternatively, the total time for brashing and/or the felling and disposal of unmeasurable trees may be worked out and divided by the number of the crop trees, the resulting figure being added to the

Standard Time per tree, and this method is to be preferred.

(e) *Moving Tips to Racks, Disposing of Lop and Top*

For moving the tips of felled trees to an extraction rack and clearing lop and top from the rack when this is done in clearfelling areas to facilitate extraction, add 0·10 SMs per tree to all trees.

(f) *Stump Treatment*

For applying nitrite, polybor or urea to the stump immediately after felling (including fetching chemicals, filling cans, etc).

Add 0·40 SMs per tree for trees up to 0·18 m³.
Add 0·60 SMs per tree for trees from 0·18 to 0·72 m³.
Add 0·80 SMs per tree for trees over 0·72 m³.

(g) *Turning Trees for Snedding*

The normal method of snedding trees which are too large to be turned from the top is to sned them as thoroughly as is possible without turning them. When it is considered necessary to turn such trees from the butt using a cant-hook add 5% to the felling Standard Times.

(h) *Turning Timber for Snedding*

When trees are crosscut at stump into one or more pieces of timber and a pole it is usual to turn the pieces during snedding. Normally this work is covered by the Standard Times, but in difficult conditions or when trees with a volume of more than 0·75 m³ or pieces of more than 0·25 m³ are being crosscut an addition to the felling times may be necessary. When in such conditions it is considered essential to turn the **timber** to ensure an adequate standard of snedding. *Add 5% to the felling Standard Times.*

(i) *Trees Snedded to Top Diameters less than 7 cm*

When cutters are required to sned poles beyond 7 cm top diameter additions should be made to the Standard Times as shown in the table below. The extra time for snedding to the average cut-off diameter should be added to all trees.

Snedding from 7 cm top diameter to	*Standard time per tree in minutes*
6 cm top diameter	0·31
5 cm top diameter	0·58
4 cm top diameter	0·84
3 cm top diameter	1·08

N.B. These times should not be used for trees averaging more than 0·20 m³ in volume to 7 cm.

Drift

Standard times are subject to a phenomenon termed 'drift'. In other words, as time passes minor changes take place each of which taken individually is unlikely to be observable using a stopwatch, but which together have a significant effect. The workers themselves find slightly better ways of doing the job, chainsaw manufacturers make minor improvements to their products, more training is given, etc. The overall effect is that the times become slack. The time study man can carry out check studies at intervals to determine the extent of drift and generally speaking it is usual to restudy major Standard Time Tables at three-yearly intervals.

Output Guides

Time study is expensive and the degree of study necessary to produce a statistically validated table of standard times for a major operation (such as thinning Sitka spruce) cannot be justified for an operation such as peeling by a Cambio peeler as so few of these machines are in use in the United Kingdom. Tables with a lower precision, called Output Guides, are produced in these instances. These Output Guides nevertheless are of considerable value to forest managers and to more senior staff not only for payment purposes but also for planning. They are not usually subjected to the same degree of restudy as Standard Time Tables are.

Output Guides are produced:

(a) When for any reason it has not been possible to gather sufficient data to enable a full Standard Time Table to be produced.

(b) When the expense of producing a full Standard Time Table is not justified because of the local* nature of the work or because the work is a minor operation in the country as a whole.

(c) When guidance is required by management quickly (ie before the full studies can be completed) for example, where early guidance is needed on times for a new working method. After further studies such a guidance table may be replaced by a full Standard Time Table.

(d) When data collected in one area (and perhaps published as a Standard Time Table) is modified by Work Study Branch for use in another area without a full programme of studies.

*Note: Output Guides of purely local significance are not reproduced in this publication.

Output Guides are similar to Standard Time Tables but para 6 may give the likely output per hour in terms of quantity of work achievable instead of the time for the job.

All Output Guides have standard headings:

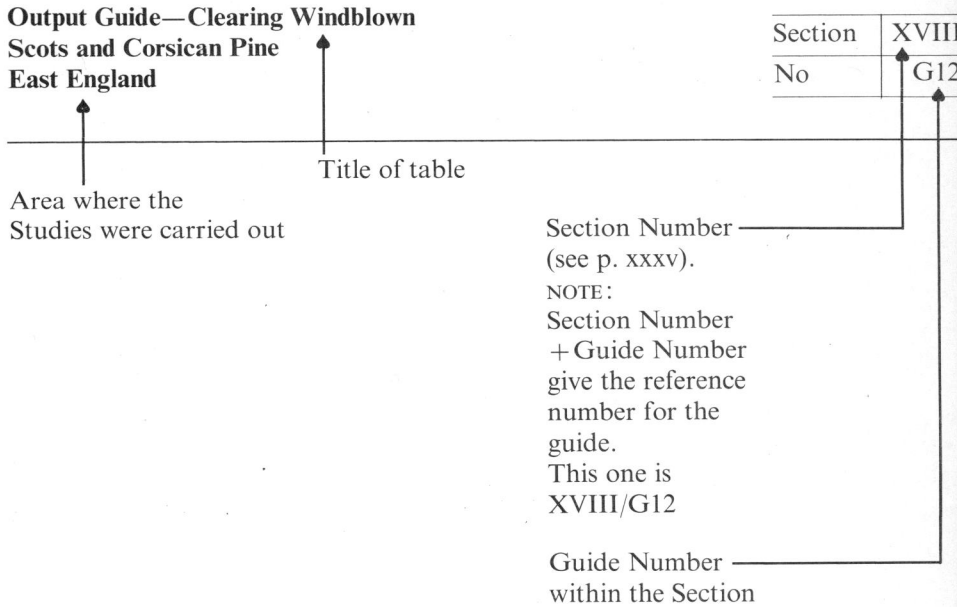

Output Guide—Clearing Windblown
Scots and Corsican Pine
East England

Section	XVIII
No	G12

Title of table

Area where the
Studies were carried out

Section Number
(see p. xxxv).
NOTE:
Section Number
+ Guide Number
give the reference
number for the
guide.
This one is
XVIII/G12

Guide Number
within the Section

The paragraph numbering system is the same as for Standard Time Tables but para 6 sometimes gives the time for the job (as with a Standard Time Table) or it gives the output per hour or per day. An example of the latter for step and notch planting (Output Guide No XIV/G1) is:

Outputs

Site	Slope		Output Trees/Hour
	Degrees	*%*	
A	0–5	0–9	235
	6–10	10–18	220
	11–16	19–29	206
	17–21	30–38	194
B			167

Uses of Standard Time

Standard Time Tables can be used for the following purposes each of which is considered below:

(i) To form the basis of an incentive payment scheme.
(ii) To calculate output.
(iii) To assess performance.
(iv) To calculate labour requirements.
(v) To calculate machine requirements.
(vi) To calculate productivity increases.
(vii) To cost an operation.

From the fact that Output Guides are usually produced from fewer studies than are Standard Time Tables, it follows that less reliance can be placed on their accuracy. Nevertheless, with the proviso that they may be less accurate, they can be used for the same purposes as can full Standard Time Tables. In the remainder of this publication, the term 'Standard Time Table' should be read as including 'Output Guides'.

Converting Time to Rate for the Job
In its simplest terms:

Time × money = Rate for the job.

If we therefore select the time for the job from para 6 (as modified by para 7) of the appropriate table, we need only multiply it by an agreed price for a minute of work and we have the appropriate rate for that unit of work.

Calculating the Value of a Minute of Work—The Price per Standard Minute
For illustration purposes, let us assume a weekly rate of pay for a skilled forest worker of £40 for a basic 40 hour week. The value of one minute of work is therefore $\dfrac{£40}{40 \times 60} = £0 \cdot 0167$ or $1 \cdot 67$p.

However, if a worker is paid 1·67p for each minute of work done and he were an average skilled man, he would earn £40 per week but would give only day-work output because there is no incentive for him to do otherwise.

It is usual therefore to negotiate an incentive allowance which may vary depending on the degree of skill required to do the job, for example, to apply herbicides accurately, to operate a processor. Let us assume that an incentive allowance of 30% has been agreed by men, management and unions. Then:

$$\text{Price per Standard Minute} = \frac{\text{Basic Weekly pay in pence}}{40 \text{ hours} \times 60 \text{ minutes}} \times \frac{130}{100}$$

$$= \frac{£40 \times 100}{40 \times 60} \times \frac{130}{100} = 2\cdot17\text{p}$$

The average worker can now earn at least day pay plus 30%, that is £40 + 30% = £52 per week. But this is not a ceiling to earnings; if the man rates more than 100, then he will earn more. For example, a man who rates 125 will be able to earn around £65 per week which is the equivalent of day pay + $62\frac{1}{2}\%$. If he takes less rest than is allowed in the Standard Time Table his pay can go up to the day pay + 80% level.

Calculating the Rate for the Job—Example

If we find, by looking at the appropriate table XV/G10 that the time for weeding one hectare of woody weeds in 2·1 m × 2·1 m spacing using a portable brushcutter is 1,040 standard minutes, then the rate for the job is:

1,040 × 2·17p = £22·57 per hectare.

Calculating Output

If the Standard Time for a job, say felling and snedding a tree, is ten standard minutes, then the output for an eight-hour day at standard performance (ie for a man rating 100) is:

$$\frac{\text{Minutes at work daily}}{\text{Standard minutes per tree}} = \frac{8 \text{ hours} \times 60 \text{ minutes}}{10 \text{ standard minutes}} = 48 \text{ trees per day}$$

A man will not in fact take ten minutes at standard performance to fell and sned a tree for which the Standard Time is ten minutes, but perhaps six-and-a-half minutes; the balance is other work and rest taken at different times during the day.

If the operator consistently works at more than 100 rating, his output will be higher. For example: if in the above job the operator rates 110, his output per day will be $\frac{48 \times 110}{100} = 53$ trees approximately.

The output as calculated above is for one man; if as part of the specified conditions of the job, two or more men are required to work together as on some machines, then the output for that machine is:

$$\frac{\text{Minutes at work daily}}{\text{Standard minutes for job}} \times \text{Number of operators.}$$

Calculating Performance

The actual output of a less skilled or day worker and highly skilled pieceworker will differ considerably and the output of different workers felling trees with a standard time of ten minutes might well range from 36 to 60 trees per day. Standard performance (ie that achieved by a skilled worker working under incentive conditions and rating 100) may be expressed as 100. The output given above can then be expressed relative to this and the performance calculated:

Output achieved	Output at 100 Performance	Actual Performance
36 trees	48 trees	$\frac{36}{48} \times 100 = 75$
60 trees	48 trees	$\frac{60}{48} \times 100 = 125$

Expressing these calculations in general terms, performance can be defined as:

$$\frac{\text{Output expressed in standard time}}{\text{Time on job}} \times 100$$

eg with an output of 55 trees of ten standard minutes each in an eight-hour day and one-man working:

$$\frac{55 \text{ trees} \times 10 \text{ standard minutes}}{8 \text{ hours} \times 60 \text{ minutes}} \times 100 = \text{a performance of 115.}$$

In forestry, more than in manufacturing industries, performances over short periods may appear to vary widely when in fact it is local variations in conditions which are responsible. However, performances over periods of more than a few days which consistently fall outside the limits of 75–130 need investigation, because the job or conditions may have changed or the standard data may have been incorrectly applied.

It should be noted that a performance of 75 represents approximately day-work output, while for a performance of 130 to be maintained all day even by an exceptionally good pieceworker is unusual. Management is of course free to investigate lesser variations; and on the other hand, it should not be overlooked that apparent performances outside the 75–130 limits may be due to more or less rest being taken than is specified in the Standard Time Table.

Calculating Labour Requirements

In order to calculate the labour requirement for a job it is first necessary to obtain from the Standard Time Table the total time required for the job. If this total is then divided by the time in minutes for which one man may be expected to work during the period available, the result is the number of workers required.

Example: To fell and sned 4,000 trees (of a given species) average volume $0.55\,m^3$ in a period of four weeks (each of five days). How many men are needed?

Let us assume that the time per tree from the appropriate Standard Time Table is 14·40 standard minutes.

Therefore, the time for the stand is $14.40 \times 4,000 = 57,600$ standard minutes. As there are 2,400 minutes per man per week available, the number of men required is $\dfrac{57,600}{2,400 \times 4} = 6$ men.

This gives a tight schedule and assumes no lost time on account of bad weather or other unforeseen stoppages. It also assumes that the men are working at standard performance. In practice, they may well exceed standard performance finishing the job early, and perhaps making up for modest periods of lost time.

Calculating Machine Requirements

The calculation of machine requirements is similar to that for labour but must allow for the size of the team working the machine. Example: A small depot has a requirement for a machine to peel $12,000\,m^3$ of poles averaging $0.09\,m^3$ each and $8\,m$ long per annum. Limited time study of a new peeler has produced a standard time of 18 minutes per 100 m peeled with a crew of two men. Would this machine meet the depot's needs without undue overtime or shift working?

Bearing in mind that the standard time is for a crew of two, then the time for the machine per 100 m is $\dfrac{18}{2} = 9$ minutes.

Total length of poles to be peeled $= \dfrac{12,000}{0.09} \times 8\,m = 1,066,667\,m.$

Therefore number of days required $\dfrac{1,066,667 \times 9}{480 \times 100\,m} = 200$ days.

Two hundred days, for all practical purposes, represent about a year's work but it will probably be necessary to introduce a little overtime for machine maintenance and unforeseen holdups. On the evidence gleaned from the limited studies done, the manager will probably accept that the machine will meet his requirements.

Calculating Productivity Increases

One of the advantages of long-term work study is that it produces data that can be stored and recalled in the years to come. It can be stored together with a short cine film showing precisely what the correct method of working is. This store is known as a 'data bank'.

If we wish to know how productivity has increased over say the last 18 years in felling a certain species of tree of given size, then as long as we have stored the 1960 data, there is no problem.

Let us 'retrieve' the time for man to fell trees averaging $0.2\,m^3$ using a bowsaw and an axe (the latter for snedding) in 1960 and compare it with today's (1978) time for chainsaw felling and snedding:

	Standard Minutes per Tree	Approximate Number of Trees per day (480 mins)
1960	15·60	31
1978	6·72	71

Time saved
over 18 year period $= 8.88$ Increased output $= 40$ trees per day
With an extra 40 trees being felled per day, productivity has increased over 18 years by 129%. By using discounting tables (as used by the insurance profession) we can work out that the average productivity increase per annum has been 4·7%.

Costing an Operation

The four important factors to be taken into account when costing an operation are:
* Labour Cost
* Labour Overheads
* Machine Charges
* Materials Cost

Any costing exercise should state clearly that all of these factors, as appropriate, have been included.

By labour overheads, is meant those expenses (such as holidays, sick pay, national insurance, pension contributions, wet time, worker transportation, etc) which are there because the man is employed and which are not there if he resigns or is dismissed.

The hub of the costing system is the standard time for the job. Each of these four factors can be converted to a price per minute of work; let us look at them individually:

Labour

This has been dealt with on p. xxi. Refreshing your memory: to calculate a price per standard minute of work, the week's basic pay is divided by the number of minutes in a week and multiplied by the agreed incentive, for example:

$$\frac{£40 \times 100}{40 \times 60} \times \frac{130}{100} = 2\cdot17 \text{ pence per standard minute (p/SM).}$$

Labour Overheads

The total cost of the items which together comprise labour overheads is a closely guarded secret in most companies as it is so easily used as a dumping ground for 'miscellaneous overheads' which often reflect the firm's efficiency (or inefficiency). This example therefore is based on figures 'grabbed from the air'! Suppose that last year a certain company employed over the year an average of 100 men and incurred total labour overhead charges of £48,000. Allowing for wet time, sickness, paid holidays, etc, a man works about 200 days per year in the woods.

Therefore the average yearly labour overhead per worker in the company is £480. Per day, this works out at £2·40 and can also be expressed as a charge per minute of $\frac{£2\cdot40}{480} = 0\cdot50$p. This is of course really a means of distributing the *current* year's costs on the basis of what happened *last* year. It may be advisable to increase the charge to cater for foreseen increases in the current year.

Total Labour Charges

It is very easy to make the mistake of thinking that the cost of a worker is the wage he is paid expressed perhaps as a rate for the job. Though a man may be paid 2·17p/SM while he is felling trees, the extra 0·5p per minute must not be overlooked when costing the job.

Machine Charge

Machines can be charged on a similar basis to labour overheads; if your records show that a certain type of tractor cost £1·00 per hour to run last year (for such things as fuel, lubricants, repairs, maintenance, insurance, depreciation, etc, but excluding operator), you could make an estimate of this year's likely cost, say £1·20 per hour. The price per minute therefore would be 2·0p.

Using historical costs has severe disadvantages however; a machine may not have been used very much, as for example, a fire standby vehicle; it may have been involved in an accident and had a high incidence of repairs; it may be new and produce low costs, or old and produce high costs. You may be buying or contemplating buying a machine of a type you have not used before and therefore have no historical costs to call upon. In these cases a good *estimate* of the cost of running a machine can be obtained from this formula:

$$\frac{\text{Capital Cost of Machine} \times 3}{\text{Estimated Life in Hours}} = \text{Cost per hour}$$

Let us take as an example a forwarder costing £18,000; we expect it to work effectively for 200 days a year for 5 years. The sum then becomes:

$$\frac{£18,000 \times 3}{5 \times 200 \times 8} = £6{\cdot}75 \text{ per hour}$$

Expressed as a price per minute, the forwarder would be charged at 11·25p/SM. (A useful 'rule-of-thumb' guide is to remember that with a life of 8,000 hours, under this formula each £1,000 of capital costs 37·5p per hour or 0·625p per minute).

Taking as another example, a chainsaw costing £150 (including spare bars and chains and less trade-in value on sale) which, used on felling and snedding works $6\frac{1}{2}$ hours per day for a year and is actually running for 60% of the time, then the sum becomes:

$$\frac{£150 \times 3 \times 60\%}{6\frac{1}{2} \times 200 \times 100} = 21 \text{ pence per hour or } 0{\cdot}35\text{p per minute}$$

If the worker owns the saw then a price per hour *in use* should be agreed with him. This sum can either be paid for separately or added to his price per standard minute. Eg if the negotiated rate per hour in use is 40p, and the saw is used on felling and snedding, the saw will cost $\frac{40}{60} \times 60\% = 0{\cdot}40\text{p}$ per minute, the worker 2·17p. Hence, the man who fells and sneds trees with a company saw is paid 2·17p/SM and the man with his own saw 2·17 + 0·40 = 2·57p/SM.

Costing a Job—Example

Let us go back to the example used earlier (p. xxiv) of how many men would be needed to fell and sned 4,000 trees (of a given species) average volume 0·55m³ in a period of 4 weeks (each of 5 days). Let us add a supplementary question now—'and how much will it cost?'

Total standard time = 57,600 minutes

		% of cost
Rate per man per minute	2·17p/SM	71·9
Rate for labour overheads per minute	0·50p/SM	16·5
Rate for saw per minute	0·35p/SM	11·6
Total price per standard minute	3·02p/SM	100

The percentage figures given demonstrate just how labour intensive felling and snedding a tree still is!

The total cost of the job is $\dfrac{57,600 \times 3 \cdot 02p}{100} = £1,739 \cdot 52$, say £1,740.

To this can be added the desired profit expected and then a quotation can be given.

Where materials are used, as in chemical spraying, fertilising or fencing, they can be added in at this stage.

In this example, the cost per tree is $\dfrac{£1,740}{4,000} = 43 \cdot 5p$, and the cost per m³ is $\dfrac{£1,740}{4,000 \times 0 \cdot 55} = 79 \cdot 1p$.

This technique of costing not only enables you to give meaningful estimates, but also to check those you receive from firms contracting to you.

Comparing Methods of Working

It is all too easy to overlook some important factor when costing an operation. If we want to know whether it is cheaper to weed by hand or by portable brushcutter, we can delude ourselves by forgetting the machine charge (say 50p per hour).

For example, we need to weed an area of 20 hectares of brambles in $2 \cdot 1 \, m \times 2 \cdot 1 \, m$ spacing. The time per hectare by hand is 1,190 standard minutes and by portable brushcutter 960 standard minutes. Which method do we use?

Comparison

	Hand	Portable Brushcutter
Man	$1,190 \times 2 \cdot 17p = £25 \cdot 82$	$960 \times 2 \cdot 17p = £20 \cdot 83$
Labour overheads	$1,190 \times 0 \cdot 50p = £5 \cdot 95$	$960 \times 0 \cdot 50p = £4 \cdot 80$
	Total £31·77 per ha	Total £25·63 per ha

The machine appears to save £6·14 per hectare but we have forgotten the machine cost at 50p per hour or 0·83p/SM. Adding this to the cost of the portable brushcutter ($960 \times 0 \cdot 83p = £7 \cdot 97$), the true brushcutter cost is £25·63 + £7·97 = £33·60, thus making the brushcutter £1·83 per hectare dearer than the hook.

But the decision cannot be taken on direct cost alone, the job must be viewed as a whole. If for example the area to be weeded were 0·5 hectare, a man with a portable brushcutter would weed it in one day (960 SMs $\times$ 0·5 = 480 minutes = 1 day) whereas a man with a reap hook would take almost a day and a quarter thus requiring two visits to the site. The saving of 92p by using the reap hook is more than lost by making the two journeys.

Daywork or Piecework?
Probably not much work is carried out on straight daywork nowadays but it is important to know the cash advantages (not just to the worker but to his employer too) of piecework. Piecework assumes a higher output with no increase in labour overheads. Let us compare weeding on daywork and on piecework.

Daywork

Output, say, 0·2 hectares per day or 1 hectare per week

5 days pay	= £40·00
Labour overheads £2·40 × 5	= £12·00
Total cost per week	£52·00
Cost per hectare	= £52·00

Piecework

5 days pay	= £40·00
Incentive pay @ 30%	= £12·00
Labour overheads £2·40 × 5	= £12·00
Total cost per week	£64·00

Output 30% above daywork = 1·3 hectares.

Therefore cost per hectare = £49·23, a saving of £2·77 per hectare over daywork.

If the worker works at a performance of 125, the situation becomes:

5 days pay	= £40·00
Incentive pay @ 30%	= £12·00
	£52·00
+25% (125 performance)	£13·00
	£65·00
Labour overheads £2·40 × 5	= £12·00
Total cost per week	£77·00

But now the output is of the order of 1·63 hectares per week. Therefore the cost per hectare is £47·24, a saving of £4·76 over daywork.

Forestry Commission (Work Study) Addresses

Work Study Headquarters
Forestry Commission
  (Work Study)
Forest Research Station
Alice Holt Lodge
Wrecclesham
Farnham
Surrey GU10 4LH
Phone: Bentley (042 04) 2255

Southern Region Team
(Responsible for England south of
the Thames and Severn and the
Forest of Dean)
Forestry Commission
  (Work Study)
Victoria Tilery
Brockenhurst
Hants SO4 7QJ
Phone: Brockenhurst (059 02)
  2227

Eastern Region Team
(Responsible for England south of
Manchester and north of the
Thames and Severn)
Forestry Commission
  (Work Study)
c/o District Office
Santon Downham
Brandon
Suffolk IP27 0TJ
Phone: Thetford (0842) 810271

Wales Team
(Responsible for whole of Wales)
Main Office:
Forestry Commission
  (Work Study)
Boughrood House
The Struet
Brecon
Powys LD3 7LS
Phone: Brecon (0874) 2557

Sub Office:
Forestry Commission
  (Work Study)
Agriculture House
Lion Street
Dolgellau
Gwynedd LL40 1DH
Phone: Dolgellau (0341) 422289

Borders Region Team
(Responsible for Scotland south of
Edinburgh/Glasgow and England
north of Manchester)
Main Office:
Forestry Commission
  (Work Study)
Kielder Castle
Hexham
Northumberland
Phone: Kielder (0660) 50235

Sub Office:
Forestry Commission
  (Work Study)
Stable Cottages
Mabie
Dumfries DG2 8HB
Phone: Dumfries 2762

Northern Region Team
(Responsible for Scotland north of
Edinburgh/Glasgow)
  Forestry Commission
    (Work Study)
  Millyard
  Smithton
  Inverness IV1 2NL
  Phone: Inverness 39010

Forestry Commission Publications mentioned in this Booklet

Leaflet 55 (1973)   **Hydratongs**   F. B. W. Platt and P. Wood
HMSO
Leaflet 61 (1975)   **Tubed Seedlings**   A. J. Low and J. S. Oakley
HMSO
Leaflet 62 (1975)   **Ultra Low Volume Spraying**   E. V. Rogers
HMSO
Leaflet 75 (1978)   **Harvesting of Windblown Trees**   H. E. Jones
and R. O. Smith
Booklet 36 (1973)   **Timber Measurement for Standing Sales using
Tariff Tables**   G. J. Hamilton   HMSO
Bulletin 48 (1974)   **Weeding in the Forest**   W. O. Wittering
HMSO
Entopath News Chemical Control Supplement   Forestry
Commission   O. N. Blatchford (Editor)   Available free of
charge from the Principal Entomologist, Forestry Commission,
Alice Holt Lodge, Wrecclesham, Farnham, Surrey GU10 4LH

List of Forest Operations

List of Standard Time Tables
and Output Guides

Ploughing for Planting
North, West & South Scotland

1 Conditions

The Guide applies to ploughing on three clearly defined treatment types. The treatment types control the type of equipment which is used. (See paragraph 3.)

The three treatment types are as follows:

a. *Treatment type 1*   Peat areas requiring aeration and local drainage with peat over 0·3 m deep.
Vegetation: Molinia and/or sphagnum and/or eriophorum.
Ranging in fertility.
Slope normally less than 12% (7°).
Surface obstructions few apart from some drains and old 'peat hags'.
Underground obstructions: Roots and mineral patches.

b. *Treatment type 2*   Mineral sites requiring cultivation by trailed plough.
Peat usually less than 0·3 m deep.
Vegetation: Calluna and/or scirpus and/or vaccinium and/or grasses.
Ranging in fertility.
Slope generally over 12% (7°).
Obstructions common both above and below ground.
Possible to turn at furrow ends.

c. *Treatment type 3*   Mineral sites requiring cultivation by mounted ploughs as type 2 but:
Slope 27% (15°) or over.
Difficult or impossible to turn at the top of the furrow.

2 Job specification

The outputs are for the following work:

a. Ploughing at 2 m intervals to the specification of the ploughs, which are:
1. Tine ploughs—tine depth 45–55 cm.
2. Turfing ploughs—36–51 cm.

1

b. The outputs include all travelling necessary during the course of ploughing, turning at the end of furrows, travelling light from the end of one furrow to the start of the next, lifting and crossing ride lines etc, lifting to free the plough from boulders or roots, and small delays due to the nature of the ground provided they are less than 15 minutes duration.

c. The Guide includes those small items of maintenance which are generally by nature an adjustment of the equipment, such as setting discs, changing sock in the field etc. It also includes an allowance for the routine half hour maintenance per day.

3 Tools and equipment (including safety equipment)

a. The tractors and ploughs used will depend on the treatment type and the Output Guide should only be used for the following combinations:

Treatment type 1
Tractors: Bowen 60 (Mark I and II), TM 55 and 70, International 500.
Plough: Universal Turf Single Mouldboard.

Treatment type 2
Tractors: Fiat 70, International BTD 8 (occasionally TM 55 and 70).
Plough: Universal Tine Single Mouldboard.

Treatment type 3
Tractors: Fiat 70, International BTD 8.
Plough: Mounted Tine Single Mouldboard.

b. All tractors and ploughs should be equipped with tools for maintenance and spare bolts, socks, discs, etc.

c. Protective clothing viz:
Safety helmet BS 2826 or 5240.
Boots with safety toecap.
Ear defenders.

4 Allowances included in the outputs
The Guide has been produced from gross data and consequently it includes allowances for minor operations which inevitably occur and for rests.

5 Method of using the Output Guide

a. Decide into which treatment type the area falls.

b. Using the appropriate table 5A, 5B or 5C consider each characteristic in turn and decide which factor is the most appropriate to the site in question.

c. Note the score corresponding to that factor.

d. Having considered all characteristics, total the score.

e. The output in hectares per hour may then be read from the table in paragraph 6 using the total score in column 1 and the appropriate treatment type in columns 2 to 5.

Example

North Scotland.

Treatment Type One.

Job Specification and Tools and Equipment as shown in paragraphs 2 and 3.

Length of Furrow = 230 m.

Bogging risk = Medium.

Slope not exceeding 10% (6°) but with rough broken surface.

Peat with mineral soil common.

All 1-way ploughing.

Total Score = 2 + 4 + 4 + 3 (From table 5A).

= 13.

Output = 0·16 ha per hour (From table 6A).

5A—*Treatment type 1*

Characteristic	Factor	Score
Length of furrow	a. Over 400 m	1
	b. 150–400 m	2
	c. Under 150 m	3
Bogging risk	a. Low	2
	b. Medium	4
	c. High	6
Slope and surface	a. Flat/gentle slope with gentle undulations	2
	b. Slope not exceeding 10% (6°) but with rough broken surface— 'ruts' and/or 'tussocks' present	4
	c. Slope may exceed 10% (6°) and surface very rough and broken, crossing old drains, peat 'hags' etc	6
Plough obstructions (underground)	a. Pure peat with only occasional roots	1
	b. Peat with roots or patches of mineral soil	2
	c. Peat with mineral soil and/or roots common	3

Characteristic	Factor	Score
Length of furrow	a. Over 400 m	1
	b. 150–400 m	2
	c. Under 150 m	3
Tractor obstructions (surface)	a. Few or absent. Cause no trouble to the equipment	1
	b. Occasional rocks or boulders requiring negotiation	2
	c. Frequent boulders requiring action to avoid. Often causes the tractor to slow down	3
Plough obstructions (underground)	a. Few. Cause no slowing	1
	b. Occasional lifting of plough necessary	2
	c. Frequent lifting and/or slowing of the tractor	3
Travel light/turn	a. No difficulty. Choice of routes available. Turning at furrow ends no problem	1
	b. Some difficulty. Routes restricted. Turning may present the occasional difficulty	2
	c. Difficult. Route may be indirect. Turning difficult as a rule.	3
Slope and surface	a. Causes little trouble, either the slope is slight or the going is firm	1
	b. Combination of slope and surface is such that care has to be taken at some point	2
	c. Conditions difficult, tractor in danger of losing grip. Slow driving necessary	3

Characteristic	*Factor*	*Score*
Length of furrow	a. Over 250 m	1
	b. 100–250 m	2
	c. Under 100 m	3
Tractor obstructions (surface)	a. Few or absent. Cause no trouble to the equipment	1
	b. Occasional rocks and boulders requiring negotiation	2
	c. Frequent boulders requiring action to avoid. Often causes the tractor to slow down	3
Plough obstructions (underground)	a. Few. Causing no slowing	1
	b. Occasional lifting of the plough necessary	2
	c. Frequent lifting and/or slowing of the tractor	3
Travel light/turn	a. No difficulty. Choice of routes available. Forward travel possible	1
	b. Reversing alongside previous furrow	2
	c. Reversing difficult. Danger of slipping. Boulders present	3
Slope and surface	a. Causes little trouble, either the slope is gentle or the going is firm	1
	b. Combination of slope and surface is such that care has to be taken at some point	2
	c. Conditions difficult. Tractor in danger of losing grip. Slow driving necessary	3

6 Scotland outputs

6A *North Scotland outputs*
Output in hectares per hour

Score	Treatment Types				
	1 *(all 1-way)*	*1* *(all 2-way)*	*2*	*2* *(all 2-way)*	*3*
5	—	—	0·21	—	0·19
6	0·23	0·40	0·20	—	0·18
7	0·22	0·37	0·19	0·40	0·17
8	0·21	0·34	0·17–0·18	0·35–0·36	0·16
9	0·20	0·31	0·16	0·31	0·15
10	0·19	0·28	0·15	0·26–0·27	0·14
11	0·18	0·25	0·14	0·22	0·13
12	0·17	0·22	0·13	0·17–0·18	0·12
13	0·16	—	0·12	—	0·11
14	0·15	—	0·10–0·11	—	0·10
15	0·14	—	0·09–0·10	—	0·09
16	0·13	—	—	—	—
17	0·12	—	—	—	—
18	0·11	—	—	—	—

6B   *West Scotland outputs*
Output in hectares per hour

Score	Treatment Types			
	1 *(all 1-way)*	*1* *(all 2-way)*	*2* *(all 1-way)*	*3*
5	—	—	0·26	0·29
6	0·27	0·38	0·25	0·275
7	0·265	0·37	0·24	0·255
8	0·26	0·35	0·23	0·235
9	0·25	0·34	0·22	0·22
10	0·24	0·33	0·21	0·205
11	0·235	0·315	0·20	0·185
12	0·23	0·30	0·19	0·165
13	0·22	0·29	0·18	0·15
14	0·21	0·28	0·17	0·135
15	0·20	0·26	0·16	0·12
16	0·195	0·25	—	—
17	0·19	0·24	—	—
18	0·18	0·22	—	—

6C *South Scotland outputs*
Output in hectares per hour

Score	Treatment Types				
	1 *(all 1-way)*	*1* *(all 2-way)*	*2* *(all 1-way)*	*2* *(all 2-way)*	*3*
5	—	—	0·24	No data	0·29
6	0·265	0·425	0·23	No data	0·27
7	0·25	0·405	0·215	0·415	0·25
8	0·24	0·385	0·20	0·38	0·23
9	0·225	0·365	0·19	0·33	0·21
10	0·215	0·345	0·18	0·285	0·19
11	0·205	0·325	0·165	0·245	0·17
12	0·19	0·305	0·15	No data	0·15
13	0·18	0·285	0·14	No data	0·13
14	0·165	0·265	0·13	No data	0·11
15	0·155	0·245	0·115	No data	0·09
16	0·145	0·225	—	—	—
17	0·13	0·205	—	—	—
18	0·12	0·185	—	—	—

7 Modifications and variations to the outputs

N.B. These modifications and variations are only to be applied when the prevailing conditions or job specification differ from those listed in paragraphs 1 and 2.

a. *Twin furrow plough*

This plough gives complete cultivation. It is pulled by a Challenger 33 and experience to date on easy sites (score 6–7·5 according to the factors in table 5B) gives outputs:

1-way ploughing—0·22 ha/hour
2-way ploughing—0·25 ha/hour

Drains Maintenance

Section	XIII
No	G1

---

1 Conditions

The times apply to deepening or deepening and widening peat drains with either peat or mineral soil bottoms, and for deepening drains in mineral soil under the following conditions:

a. *Vegetation*

    (i) Mainly molinia, rushes, polytrichum, sphagnum, bilberry and heather with occasional gorse and broom,

    (ii) Mainly heather with eriophorum, molinia, sphagnum and mosses,

or (iii) Mainly grasses, rushes and mosses, with some heather.

b. *Soil*

    (i) Peat with occasional large stones and silt,

    (ii) Clay and silt with few stones,

    (iii) Clay and silt mixed with stones,

or (iv) Hard packed clay and silt with many stones and shale.

c. *Roots*

    (i) Heather roots and occasional roots of live or dead trees requiring cutting.

2 Job specification

The times are for the following work:

a. Cutting vegetation overhanging the drain, widening and deepening the drain to the specification required by the supervisor.

b. Finishing the drain with a bottom width of not less than 23 cms.

c. Placing spoil clear of the drainside on the downhill side of the drain whenever possible.

3 Tools and equipment (including safety equipment)

a. Rutter spade (or side cutter).

b. Drag (hack) 3-pronged.

9

c. Bottoming spade (or bottoming shovel).

d. Flat file and handle.

e. Protective clothing viz:

Eye protection (if brashing necessary).
Boots with safety toecap.

Additional tools to be available as necessary

f. Slasher.

g. Reap hook.

h. Brashing saw and maintenance equipment.

i. Pick (chisel and point).

j. Mattock (grubbing).

4 Allowances

Included in the times:

a. For contingencies and work other than actual draining, eg sharpening and cleaning tools, walking to the next drain etc. 10% of the time spent on drains maintenance.

b. For personal needs and rest according to the type and depth of soil removed 20 to 29% of the total working time.

5 Method of using the Output Guide

Sufficient samples of each type of drain to be cleaned should be measured to ensure adequate coverage of the area.

a. *Depth*

Measure the vertical depth of the drain from ground level compressing the surface vegetation and any vegetation in the base of the drain by foot pressure. Subtract the average depth measured from the depth required and read the time from Table 6A according to the nature of the soil: if there is a mixture of peat and mineral soil measure the depth of each, read the times from Table 6A and add them together. The work of deepening includes normal side cutting and cleaning and in well-maintained drains, when this is all that is necessary, only the depth need be measured and the times may be read from Table 6A.

b. *Width*

In peat drains where the sides have closed in, in neglected drains, and in drains which were originally dug too narrow, widening is necessary. Measure the width of the drain at ground level, compressing the vegetation overhang. Subtract the average width measured from the width required at finish and read the time from Table 6B.

c. **When drains are both deepened and widened, add times for deepening (Table 6A) and widening (Table 6B).**

Example

Vegetation—Mainly heather with eriophorum, molinia, sphagnum
and mosses.

Soil   —Hard packed clay and silt with many stones and shale.

Roots   —As shown in paragraph one.

Job specification   —As shown in paragraph 2.

Tools and equipment—As shown in paragraph 3.

Depth removed   —$22\frac{1}{2}$cms.

Time per metre for deepening = 5·8 SMs (from para 6A).

6 Times per metre for deepening and widening drains

6A   *Time in minutes per metre for deepening drains*

Soil type	Depth removed cm										
	5	$7\frac{1}{2}$	10	$12\frac{1}{2}$	15	$17\frac{1}{2}$	20	$22\frac{1}{2}$	25	$27\frac{1}{2}$	30
(i) Peat											
Peat and silt or silt	1·5	1·7	1·9	2·1	2·3	2·5	2·7	3·0	3·2	3·4	3·6
(ii) Clay with few stones	1·9	2·2	2·5	2·8	3·0	3·3	3·6	3·8	4·1	4·4	4·7
(iii) Clay mixed with stones	2·4	2·7	3·1	3·4	3·8	4·1	4·5	4·8	5·2	5·5	5·8
(iv) Hard packed clay with many stones and shale	2·9	3·3	3·7	4·1	4·5	5·0	5·4	5·8	6·3	6·7	7·1

The peat and silt table (i) may be extended (at 0·2 minutes per metre
for each extra 2·5 cm of depth) to a maximum of 60 cm depth
removed.

6B   *Time in minutes per metre for widening drains with peat sides*

Width increase at top of drain (cm)	Up to 15 cm	15 to 30 cm	30 to 45 cm	45 to 60 cm
Minutes per metre	1·8	2·0	2·2	2·3

7 Modifications and variations to the times

N.B. These modifications and variations are only to be applied
when the prevailing conditions or job specification differ from those
listed in paragraphs 1 and 2.

a. *Increased width at the foot of the drain*

When drains are to be finished with a width of more than 23 cms
at the bottom, additions to the times for deepening drains shown
in the table should be made as follows:

28 cms bottom   *add* 5%

33 cms bottom   *add* 10%

38 cms bottom   *add* 15%

b. *Difficult widening*
   (i) When it is necessary to widen drains in young stands where, because the canopy has not closed, there is a very heavy growth of rush, heather, bilberry or tufted molinia *add* up to 15% in steps of 5% to the time for deepening and widening.
   (ii) In established plantations where widening the drain makes it necessary to cut many tree roots *add* up to 15% in steps of 5% to the time for deepening and widening.

c. *Woody weeds*
   If the growth of woody weeds, eg gorse and broom, along the drainside is dense and continuous and must be cut back with a slasher *add* up to 1·2 minutes per metre in steps of 0·3 minutes per metre.

d. *Brashing*
   (i)   Brashing sufficient only to permit access for work in small trees where branches can be cut with a slasher, *add* to the time per metre up to 0·4 standard minutes per tree brashed.
   (ii) Partial brashing of stronger branched trees requiring the use of a brashing saw but in stands where the branches are still green *add* up to 1·2 minutes per tree according to species and size of branches.
   (iii) Partial brashing or brashing in stands which have closed canopy should be calculated from the Brashing tables.

e. *Lop and top in feeder drains*
   The times include an element for removing lop and top from the drain and clearing occasional windblown trees. However, feeder drains which have become filled with lop and top after thinning and extraction will take longer. *Add* up to 0·8 minutes per metre in steps of 0·2 minutes per metre.

Erection of High Tensile Spring Steel Stock Fencing

1 Conditions
The times apply to fences erected under the following conditions:

a. On soils which enable post holes to be dug to 0·85 m.

b. On slopes up to 20° (36%) (but see paragraph 7a).

c. On fence lines which have been completely cleared of any live or dead material which might hamper the work of erection.

d. Lengths between strainers of 60–400 m. (Over 400 m see paragraph 7b.)

e. Materials to be distributed along the fence line to within a few feet of the point at which they will be used.

f. A team of two men.

2 Job specification
The times apply to the following work:

a. Posts to be fitted with a cross member, poles to be dug at least 0·85 m deep and posts to be adequately firmed.

b. Stakes to be driven in as near vertical as possible, to a depth of approximately 0·55 m.

c. All wires and netting to be strained to the correct tension and stapled at the recommended height.

d. Posts to be strutted by the recommended method.

e. Tie-downs to be fitted in low places.

f. Aprons to be fitted where tie-downs are inadequate.

g. Materials of the recommended specification to be used.

h. See Appendix II for the basic method to be followed.

3 Tools and equipment (including safety equipment)
See Appendix I.

4 Allowances included in the times
a. For contingencies and work other than that performed on actual

erection of the fence eg, to and from camp, inspect and consider, load coil on dispenser, talk to supervisor etc, 13% of the time spent on actual fence erection.

b. Personal needs and rest. Additions varying between 13 and 31% have been included in the times.

5 Method of calculating the time

a. The supervisor must make a preliminary survey of the fence line to decide and mark the position of strainers, contour, turning posts and stakes. The number of tie-downs and length of aprons should also be noted.

b. Materials are then distributed, positioning each item accurately. This is important because although the times cover essential movement of material by the fencing team, unnecessary movement due to careless distribution will make them inapplicable.

c. A time for each strain length can then be calculated as follows:

1 Divide the total length of fence line into lengths between straining posts.

2 For each 'strain length' find the number of posts, struts and other components required including tie-downs and aprons.

3 By reference to table 6A calculate the total time required for these components.

4 From table 6B find the 'Block Total' times by multiplying the block time per metre (2·5 minutes) by the strain length in metres and adding the block total per strain length (16 or 20 minutes according to the type of strainer used for straining the netting).

5 Add the total 'Component' time from table 6A to the total 'Block' time from table 6B a. and b. to obtain the total time for the 'strain' length.

d. Once times have been calculated for all the strain lengths, the time for the whole fence can be found by adding all the strain length times together. This method of obtaining the time for the whole fence is to be preferred to adding up all the component times and all the block times without obtaining individual strain length times as it makes checking easier in case of any dispute. Having obtained the time for the whole fence length one can find the average time per metre by dividing the total time by the length, and when variation between strain length is slight the fencing may be paid for at a price per metre. When, however, the change in conditions varies considerably throughout the fence line it may be better to pay for the actual strain length finished in any week in order to stabilise earnings.

Note: Walking from vehicle to the work-point on the fence line is not included.

6 Time for the erection of high tensile spring steel fencing

6A   Component times

Operation	Time in minutes			
	All soil types	Clay with few stones	0·3m of soil over shale	Peat
Dig and fill post hole, fit and firm post	—	56·0	60·0	38·0
Fix cross-member to post	9·0			
Fix one strut and thrust plate	23·0			
Bar one stake hole	—	1·1	1·2	—
Drive one stake	—	1·6	1·6	1·3
Fix tie-down (stake driven in wire stapled to stake and attached to fence)	7·0			
Fix tie-down (Molex anchor)	9·0			
Fix apron (per metre)	4·2			
Place sods to fill gaps under netting (per metre)	2·8			

6B   Block total times

a. *Block total per metre*
   Carry tools
   Fetch strainer and stakes
   Roll out and join net
   Staple wires
   Clip net   } 2·5 minutes per metre
   Staple net
   Walk and prepare
   Take out wire

b. *Block total per strain length*
   Strain wire   } (i) Using net strainers for straining net:
   Make off wire   20 minutes per strain length.
   Wrap guy links   }
   Strain net   (ii) Using monkey strainers for straining net:
   Make off net   16 minutes per strain length.

7 Modifications and variations to the times

N.B. These modifications and variations are only to be applied when the prevailing conditions or job specification differ from those listed in paragraphs 1 and 2.

a. *Slope*

On slopes with ground conditions which require abnormal care to obtain a footing,

Add 5% or 10% to the total times for the strain lengths where these conditions apply.

b. *Strain length*

For strain length of more than 400 metres,

Reduce the block total per metre obtained from table 6B a. by 5%.

Materials, tools and equipment

A *Specifications of materials*
End posts and turning posts: 2·3 metres long × 10–18 cm top
   diameter.
Stakes (pointed): 1·8 metres long × 5–8 cm top diameter.
Struts: 2 metres long × 8–10 cm top diameter.
Thrust plates (split): 0·6 metres long × 8–10 cm wide.
Staples: 40 mm long No 8 gauge.
Plain wire: High tensile spring steel 12 gauge.
Netting: Rylock hinged joint ref no C832/12.
Nails: 130 mm.
Wrap guy links: Code WGL: 0·104 inches.
Line splices: Code LS 7401 (for joining wires without using a knot).
Gordian ring fasteners.
Lashing rods: Code LRGS: 0·226 inches (of particular value where
   a stronger fastener than the Gordian is needed, eg fences liable
   to snow damage).

B *Tools and equipment*

	Number required
1. Wire dispenser	1
2. Tool belt (carpenter's belt with pouch supplied by most ironmongers or belt supplied to order by W. Lancaster and Son, 15 Little Dockray, Penrith, Cumberland)	2
3. Felco C7 wire cutters (essential for HTSS wire)	2
4. Hammer (claw type with metal handle)	2
5. Clip gun and clips (Gordian type)	2

One of each of these for each fencing belt.

6. GPO rabbitting tool with rammer modification	1
7. Shuvholer	1
8. Drivall (95 mm diameter one man) (stakes must be straight and to specification if this is to be used successfully)	1
or Stake driver, cast iron 5·90 kg (item 12 not required if this is used).	1
9. Crow bar (at least 1·4 metres long) Two crow bars are recommended where contour/fencing posts are frequent and digging is difficult.	1
10. Wire strainers (monkey)	2
11. Netting strainer	1

12. Stake driving hammer (3·18 kg) (7 lb)   1
    (If drivall is used this will be required for
    thrust plate and tie down stakes)
13. Measuring stick 1 metre long   1
14. Sandvik saw   1
15. Hatchet or small axe   1
16. Spade   1
17. Wire bending tool (supplied by Bamfords
    Limited, Uttoxeter, Staffordshire)   2

The use of radio is recommended where strain lengths and bands make communication difficult.

C *Protective clothing* viz:
Industrial gloves with wireproof palm.
Boots with safety toecap.

APPENDIX II

Method of working

The basic method to which the times apply is as follows:

1. Dig post holes. Fix cross-member to post, firm post, fix strut, thrust plate and retaining wire.
2. Take out bottom and middle wires. Keep light tension on middle wire and use to check alignment of bottom wire for use as stake driving line.
3. Drive stake and staple bottom wire at correct height. The relationship between the wire height and the ground level will then give indication of the stake spacing required. Stakes may be spaced at up to 14 metres apart where there are no undulations.
4. Take out top wire.
5. Fix tie-downs where necessary.
6. Strain wires. Because spring steel wire can be strained round a right angle bend it is not necessary to make a separate strain at every sharp change in direction.
7. Staple wires at correct height.
8. Roll out and join netting.
9. Hold netting in place sufficiently for straining by clipping to line wire where necessary.
10. Strain netting. 300 metres (6 rolls joined together) is the maximum length which should be tensioned on one straining operation.
11. Staple and clip netting.
12. Fix aprons and place sods under fence to close gaps under fence which are large enough to allow stock to enter.

Digging Drains with a Whitlock 110 Back-acting Digger Mounted on a Bowen 60 Tractor

1 Conditions

The outputs apply to draining under the following conditions:

a. Flushed basin peats with minimum depth of peat of 28 centimetres. Average drain section 124 centimetres across the top, 26 centimetres bottom width and average total depth of 87 centimetres.

b. Surface water gley. Average drain section 135 centimetres across the top, 17 centimetres bottom width and average total depth of 85 centimetres.

c. Fragio gley. Average drain section 123 centimetres across the top, 15 centimetres bottom width and average total depth of 73 centimetres.

d. Peaty weak ironpan. Average drain section 117 centimetres across the top, 17 centimetres bottom width and average total depth of 75 centimetres.

e. All drain alignments to be clearly marked for the driver.

f. Drain gradients of generally less than 3 degrees.

2 Job specification

The outputs are for the following work:

a. Dig the drain along the marked alignment to the average specification (as in paragraph 1a–d) keeping a uniform profile and a clean bottom.

b. Remove spoil well away from the drain side and deposit on the lower side of the drain.

c. A drain gradient of between $1\frac{1}{2}$ degrees and 2 degrees is desirable.

3 Tools and equipment (including safety equipment)

a. Bowen 60 tractor with a Whitlock 110 backacter digging unit plus the necessary tools for doing minor repairs and maintenance.

b. Protective clothing viz:
Safety helmet BS 2826 or 5240.
Boots with safety toecap.
Ear defenders.

4 Allowances

Included in the outputs

a. For contingencies and work other than cyclic time, eg travelling light between drains, manoeuvring, non-cyclic digging, minor repairs of not longer than 10 minutes duration, removing stumps and obstructions from the drain etc, 19% of the time spent digging.

b. For personal needs and rest 22% of the total working time.

5 Method of using the Output Guide

Paragraph 6 gives the output for the soil types specified. Modifications to these outputs can be made for changes in specification as shown in paragraph 7.

6 Output per hour

1 Deep flushed basin peat—40 metres per hour.
2 Peaty weak ironpan—39 metres per hour.
3 Gley soils (surface water and fragio)—30 metres per hour.

7 Modifications and variations to the outputs

N.B. These modifications and variations are only to be applied when the prevailing conditions or job specification differ from those listed in paragraphs 1 and 2.

a. Where drain alignments are not clearly pegged or marked for the driver, reduce the outputs as follows:

1 Deep flushed basin peat *subtract* 4 metres per hour.
2 Peaty weak ironpan *subtract* 4 metres per hour.
3 Gley soils (surface water and fragio) *subtract* 3 metres per hour.

b. On fragio gley sites completely ploughed with a mounted deep tine SMB plough—
Add 10 metres per hour to the output.

Step and Notch Planting

1 **Conditions**

    a. The outputs apply to planting by the step and notch method on single or double mouldboard ploughing in upland areas. They apply equally to tine or turf (Cuthbertson) ploughing; shallow and deep peat; regular and broken ground. See para 5.

    b. Planting of pines and spruces.

    c. Slopes up to $21°$ (38%), but see para 7b.

    d. Sheughs laid out less than 150 m average walking distance from the planting area, but see para 7a.

2 **Job specification**

The guide applies to the following work:

    a. Collecting bundles of plants from dumps, opening bundles and culling as necessary.

    b. Walking to and from dumps to planting site and within the planting site between furrows, **but not major moves between sites.**

    c. Cutting a step in the turf so as to provide shelter and to enable the roots to reach the sandwich layer—as directed by the supervisor.

    d. Planting the tree using a single notch, firming the tree and meeting the silvicultural requirements laid down.

3 **Tools and equipment** (including safety equipment)

    a. Small planting spade—Spearwell No 1152D 28″ shaft.

    b. Planting bag.

    c. File with handle for sharpening spade.

    d. Protective clothing viz:
       Boots with safety toecap.

4 **Allowances**

    a. Work other than actually cutting step and planting is 19% of stepping and planting.

21

b. Rest varies from 12%–19% between 'other work' and 'cyclic work'.

5 Selecting the output

a. Decide into which classification the site falls as follows:

A Shallow peat, peat and mineral mixture, usually associated with tine ploughing. Somewhat broken ground conditions. Slopes variable up to 21° (38%). Turf thickness less than 250 mm. Minority of plants planted in a step.

B Deep peat usually associated with Cuthbertson ploughing. Reasonably uniform conditions. Slight slope up to 5° (9%). Turf thickness over 250 mm. Majority of plants planted in a step.

b. On site type A measure the average slope.

c. Look up the outputs in para 6 below.

6 Outputs

Site	Slope		Output Trees/Hour
	Degrees	%	
A	0– 5	0– 9	235
	6–10	10–18	220
	11–16	19–29	206
	17–21	30–38	194
B			167

7 Modifications and variations to outputs

N.B. These modifications and variations are only to be applied when the prevailing conditions or job specification differ from those listed in paragraphs 1 and 2.

a. If dumps are spaced at a distance of more than 150 m from the planting area; they generally contain over 2,500–3,000 plants and under these conditions,

Reduce output by 10%

b. There is very little evidence of outputs on slopes over 21° (38%) but what little there is suggests that for every 5° (9%) increase in slope,

Reduce output by 6%

Planting Tubed Seedlings

1 Conditions
The guide applies to planting tubed seedlings under the following conditions:

a. Planting in peat where shelter has been provided in the form of a step.

b. Trays of plants are laid out throughout the planting area so that the average distance to a tray does not exceed 25 m.

2 Job specification
The guide applies to the following work:

a. Collecting full trays from the furrow ends.

b. Walking between furrows and walking to collect trays, **but not major moves between sites.**

c. Loading the planter and planting so that the tube top is below or flush with the peat surface. Tubes upright in a sheltered position. Broken or empty tubes to be replaced by sound full tubes.

d. Collect empty trays at the end of the day and pack for return to the nursery.

e. For method of working see FC Leaflet No 61.

3 Tools and equipment (including safety equipment)
a. Tubed seedling planter (see FC Leaflet No 61, page 10).

b. Carrying frame.

c. Pliers (one pair per squad usually sufficient).

d. Protective clothing viz:
Boots with safety toecap.

4 Allowances included in the guide
a. For contingencies and work other than actual planting 22% of the time spent planting.

b. For personal needs and rest 20% of the total working time.

5 Method of using the Output Guide

Site classification

The rate of planting is largely controlled by the ease of walking and the condition of the peat. Each homogeneous site is classified for peat conditions and walking conditions into one of the following classifications:

Good, medium or poor as follows:

A *Walking conditions*

Good   Little danger of stumbling or tripping. Possible to walk without concentrating on the ground. Ground reasonably firm. Possible to walk either side of the ridge easily—at least 0·3 m between ridge and furrow. Typical of good Cuthbertson type of ploughing.

Medium   A number of impediments in the way of walking so that care must be exercised at times. May be associated with tine ploughing where walking is easy on one side but very difficult on the furrow side.

Poor   Walking difficult. Care must be exercised to prevent falling. Little room to walk. Ground may be very wet.

B *Peat conditions*

Good   Peat firm, moist—but not waterlogged. Fibrous or pseudo-fibrous. Good depth of peat. Tubes penetrate easily and stay in the peat.

Medium   Peat fibrous—amorphous, occasionally hard with patches of stones and mineral soil. Typical of many tine-ploughed areas. Tubes generally penetrate easily and generally remain in the peat, but occasionally tubes may break or fail to remain in place.

Poor   Very fibrous peats with high percentage of sphagnum. Amorphous peats very friable. In both cases it is difficult to get the tube to remain in the peat. Very hard peat—dried out or with mineral surface layer making penetration difficult.

Selecting the output

Decide on the peat and walking conditions for the whole site. Allocate the appropriate factors as follows:

Classification	Peat	Walk
Good	1	2
Medium	2	4
Poor	3	6

Add the two factors together to give the 'peat and walk' factor and read off the output from paragraph 6.

Example:
Poor walking conditions, medium peat conditions.
Peat and walk factor $6+2=8$
Experienced planters
Output per hour $=450$ trees

6 Output in number of trees planted per hour

Peat and walk factor	Output per hour
3	740
4	655
5	585
6	530
7	485
8	450
9	415

7 Modifications and variations to outputs

N.B. These modifications and variations are only to be applied when the prevailing conditions or job specification differ from those listed in paragraphs 1 and 2.

a. Where this is the worker's first season of planting, the output per hour should be reduced to the following:

Peat and walk factor	Output per hour
3	730
4	625
5	545
6	485
7	440
8	400
9	365

Planting on Ploughed Ground
Notch Planting
(South Wales)

1 Conditions

The Output Guide applies to notch planting under the following conditions:

a. Planting on top of the furrows up to 30 cm in depth, formed by tine or turf ploughing.

b. Planting of pines and spruces.

c. Slopes up to 20% (11°), but see paragraph 7c.

d. Average walking distance to and from plant heel or dump of between 30 m and 160 m, but see paragraph 7b.

e. Gross areas which assume 15% unplanted ground (as rides, drains, etc), but see paragraph 7a.

2 Job specification

The Output Guide applies to the following work:

a. Collecting bundles or bags of plants from dumps, opening bundles and culling as necessary.

b. Walking to and from the dumps and within the planting site between furrows but not major moves, for example, between transport and planting site, or between planting sites.

c. Planting each tree using a single or 'L' notch 1·8 to 2·1 m apart, firming the tree with the foot and meeting the silvicultural requirements as specified by the forester, see Appendix for details.

3 Tools and equipment (including safety equipment)

a. Small planting spade, eg garden spade complying with British Standard 3388-1973 (blade length 230–290 mm, blade width 160–190 mm, handle 700 mm with Y-Dee hilt preferred) blade fitted with riveted or welded treads (or clamps to be worn on boots).

b. Planting bag—polythene net, size 46 cm × 36 cm × 15 cm.

c. File (with handle) for sharpening spade.

d. Protective clothing viz:
   Boots with safety toecap.

4 Allowances

The following allowances are included in the Output Guide:

a. For contingencies and work other than actual planting, eg collecting plants, opening bundles, culling, filling planting bags, tidying dump, sharpening spade, etc 23% of the time spent planting.
   (N.B. Major moves are excluded.)

b. For personal needs and rest 18% of the working time.

5 Method of using the Output Guide

a. Determine the predominant soil type of the turf from the following:

Turf **Soil Type A** Amorphous or pseudo-fibrous peat.
                     Mineral with a low stone content.

Turf **Soil Type B** Mineral over peat.
                     Stony mineral.

Turf **Soil Type C** Fibrous peat.
                     Stony peat.
                     Stony mineral or shale over peat.
                     Very stony mineral.

b. For the appropriate turf soil type and required plant spacing read off the time per gross hectare from the table in paragraph 6.

N.B. If it is desired to have a single time per gross hectare for an area which has a variety of turf soil types determine the percentages of each turf soil type and multiply these by the appropriate times. A total of these will give the time for the whole area, see Example 2 on page 29.

Example 1
Turf Soil Type A.
Plant Spacing 1·9 m × 1·9 m.
Slope 25% (14°).
Other conditions and job specification as shown in paragraphs 1 and 2.

Time per gross hectare (from paragraph 6)   644 SMs
Plus 5% allowance for slope (from paragraph 7c)   32 SMs
                                                  ————
                        Total time per gross hectare   676 SMs

Example 2
60% Turf Soil Type A.
25% Turf Soil Type B.
15% Turf Soil Type C.
Plant Spacing 2·0 m × 2·0 m.
Slope less than 20% (11°).
Plants well distributed (average walk of 20 m to and from dumps).
Other conditions and job specification as shown in paragraphs 1 and 2.

Time per gross hectare (from paragraph 6)

Turf Soil Type A	$582 \times 0·6$	$= 349·20$ SMs
Turf Soil Type B	$641 \times 0·25$	$= 160·25$ SMs
Turf Soil Type C	$693 \times 0·15$	$= 103·95$ SMs
Total time per gross hectare		$= 613·40$ SMs
Less 5% for good distribution (from paragraph 7b.(ii))		$= 30·67$ SMs
Net time per gross hectare		$= 582·73$ SMs

Example 3
Turf Soil Type C.
Plant Spacing 1·95 m × 1·95 m.
Slope less than 20% (11°).
Other conditions and job specification as shown in paragraphs 1 and 2.

Time per gross hectare (from paragraph 6) interpolated using formula:

$2771·85 \div (1·95 \times 1·95)$	$= 728·96$ SMs
Total time per gross hectare	$=$ say 729 SMs

6 Times per gross hectare (see para 1e) in standard minutes

Plant Spacing	Time in SMs per hectare for planting			
	1·8 m × 1·8 m	1·9 m × 1·9 m	2·0 m × 2·0 m	2·1 m × 2·1 m
Turf Soil Type A	718	644	582	528
Turf Soil Type B	791	710	641	581
Turf Soil Type C	856	768	693	629

Note: For plant spacings other than those shown in the table above, but within the limits of 1·8 m and 2·1 m, use the following formulae:

$$\text{Time per gross hectare Turf Soil Type A} = \frac{2326 \cdot 45}{\text{Plant spacing (in metres)}}$$

$$\text{Time per gross hectare Turf Soil Type B} = \frac{2563 \cdot 6}{\text{Plant spacing (in metres)}}$$

$$\text{Time per gross hectare Turf Soil Type C} = \frac{2771 \cdot 85}{\text{Plant spacing (in metres)}}$$

7 Modifications and variations to the times

N.B. These modifications and variations are only to be applied when the prevailing conditions or job specification differ from those listed in paragraphs 1 and 2.

a. *Unplanted ground*

Where the proportion of unplanted ground differs from the 15% assumed, modify the times in paragraph 6 as follows:

No unplanted ground **add** 18%
5% unplanted ground **add** 12%
10% unplanted ground **add** 6%
20% unplanted ground **deduct** 6%
25% unplanted ground **deduct** 12%

b. *Plant distribution*

(i) Where distribution is poor and the average walking distance to and from the plant dump exceeds 160 m, modify the times as follows interpolating as necessary:

Average walking distance 200 m **add** 5%
Average walking distance 300 m **add** 10%

(ii) Where distribution is good and the average walking distance to and from the plant dump is less than 30 m:

Deduct 7%

c. *Slopes*

Where slopes exceed 20% (11°) increase the times as follows:

Slopes 21%–30% (12°–17°) **add** up to 10%
Slopes 30%–40% (17°–22°) **add** up to 15%

d. *Stepping*

Where the planter cuts a step in the turf prior to planting the minority (up to 25%) of plants,

Add up to 25% to the times

e. *Single cut notches*

Where the soil type is sufficiently friable for an adequate planting hole to be made with only one cut of the spade,

Deduct 30%–40%

Working method

The normal working method to which the times in the table apply is as follows:

- (i)   Walk to next planting point estimating the distance from last planted tree.
- (ii) Cut a notch in the ploughed turf with the spade. On fibrous peat, or stony ground two cuts are assumed giving a long single notch or an 'L' notch.
- (iii) Withdraw the plant from the bag (up to six or more plants may sometimes be taken from the bag, held in the hand and planted one by one).
- (iv) Push spade (still in notch) away from body and twist slightly to open notch.
- (v)   Place plant in notch.
- (vi) Remove spade from notch and heel in the plant.
- (vii) Test firmness of plant by pulling gently.
- (viii) Repeat.

1 Conditions
The outputs apply to mechanical weeding under the following conditions:
a. Ground conditions
   (i)   slopes up to 12° (21%)
   (ii) less than 5 stumps, standing trees, or other obstacles per hectare causing delay or detour
   (iii) Ground surface providing reasonable traction.
b. Trees planted in reasonably parallel rows at 1·5 m, 1·8 m or 2·1 m row spacing.
c. Trees reasonably visible.

2 Job specification
The Output Guide applies to the following work:
a. Cut the appropriate width of weed growth (1·2 m, 1·5 m or 1·8 m) between the rows of trees at a height of 150 to 250 mm according to conditions.
b. Avoid damage to the planted crop.
c. Carry out maintenance to tractor and weeding machine as detailed in makers' hand books and in appendix I.
d. See also FC Bulletin 48, chapter 7.

3 Tools and equipment (including safety equipment)
a. Tractor and weeding machine correctly modified and guarded as appropriate to row spacing and weed growth and fitted with approved cab.
b. Tractor fuel, oil etc and all necessary tools, spares and other equipment as detailed in appendix II to be available at working site.

c. Protective clothing viz:
   Safety helmet BS 2826 or 5240.
   Ear defenders.
   Industrial gloves.
   Boots with safety toecap.

4 Allowances

a. Contingencies and other work, 20% of cutting. (This includes maintenance of tractor and weeding machine, minor repairs of not more than 30 minutes accumulated duration during the day, clearing of odd obstructions from route and crossing drains with drain crossing bridges as recommended in appendix II para 3).

b. Rest allowance 20% of total work.

5 Selecting the output

a. Ascertain the spacing between the rows.

b. Estimate the percentage cover of the four weed groups as recognised for hand weeding. (See table XV/G10). Because of the power of a machine, these can be condensed into two as follows:
   V1 covering groups A123, B123, C1 and 2 and D1.
   V2 covering groups C3 and D2 and 3.

c. Measure the average length of run the machine will make before turning.

d. Using the figures from a, b and c above look up the output from the table in para 6.

6 Output table

Output in hectares per 8 hour day.

Row Spacing	Vegetation	Average length 'weeding run' in metres						
		Up to 55	60–95	100–135	140–175	180–215	220–255	260–295
1·5m	V1	1·56	1·96	2·16	2·28	2·34	2·40	2·46
1·5m	V2	1·31	1·59	1·72	1·80	1·83	1·86	1·90
1·8m	V1	1·83	2·36	2·60	2·74	2·82	2·91	2·91
1·8m	V2	1·56	1·93	2·07	2·16	2·21	2·26	2·26
2·1m	V1	2·18	2·77	3·03	3·18	3·36	3·36	3·42
2·1m	V2	1·83	2·23	2·40	2·49	2·59	2·59	2·64

Example

a. Row spacing 1·5m.

b. Average length of weeding run 80m.

c. *Weed assessment*

Herbs A3, Grasses A2, Climbers C2   All into V1 percentage cover = 60.

Climbers C3, Woody weeds D2 All into V2 percentage cover = 40.

From output table:

Spacing 1·5 m, Vegetation V1 60%, weeding run 80 m, output

$$1·96 \times \frac{60}{100}$$

$$= 1·18 \text{ ha}$$

Spacing 1·5 m, Vegetation V2 40%, weeding run 80 m, output

$$1·59 \times \frac{40}{100}$$

$$= 0·64 \text{ ha}$$

Total output $= 1·82$ ha

7 Modifications and variations to outputs

N.B. These modifications and variations are only to be applied when the prevailing conditions or job specification differ from those listed in paragraphs 1 and 2.

a. *Obstacles*

A percentage reduction in output per 10 obstacles per hectare is necessary. An obstacle is defined as: a stump, rock, pole, standing tree from previous crop, converging row of trees, wet hole, mound, etc likely to cause delay or detour.

To make an allowance for these obstacles firstly count the number of obstacles per hectare and round to the nearest 10, then reduce the output figure obtained from the table in para 6 by the percentage obtained from the graph at appendix III, page 38.

Example

From row spacing 2·1 m, weed assessment V2, weeding run 120 m, output from the table in para 6 = 2·40 ha per day.

Obstacles per hectare to the nearest 10 is say, 30.

From the graph, using the output figure of 2·40 ha per day on the Y axis, the percentage reduction in output = 4·04% × 3 = 12·12%.

Hence the corrected output =

$$240 - \frac{(240 \times 12·12)}{100} = 2·40 - 0·29 = 2·11 \text{ ha}$$

b. *Poor visibility of trees*

If the trees are difficult to see, reduce output in steps of 5% to a maximum of 15%.

c. *Very heavy woody weed growth*

If the weed type is predominantly very heavy woody weed growth (ie it falls outside Category D3) decrease output by up to 25%.

Brushcutter maintenance

Because of the heavy strain placed on weeding machines and the vibration regular maintenance is essential to efficient working:

(i) All nuts and bolts must be checked for tightness twice daily. The use of Loctite may help to reduce trouble from this source.

(ii) The gearbox oil level must be checked daily (140 EP oil).

(iii) All greasing points must be attended to daily. (Note: on the Wolseley Swipe Jungle Buster special attention is necessary to the greasing point for cutter shaft below gearbox).

(iv) The entire machine must be inspected daily for cracks and other damage.

(v) The blades must be sharpened every one or two weeks according to weeding conditions.

Tools, spares etc required

1 *Spares for brushcutters*
The following spares should be kept with the machine:

1.1 *Blade type machines*
1 dish, 2 sets of blades, 2 pivot bolts, nuts, spring washers and keys, 1 dish retaining nut.

1.2 *Chain type machines*
1 set of chains, triangles and Allen screws.

2 *Tools for brushcutters*
The tractor tool kit should serve for normal maintenance but the following additional items are necessary:

2.1 *For Bush Hog and FES brushcutters*
$1\frac{11}{16}''$ AF socket for changing blades.
$1\frac{1}{8}''$ W socket for changing dish.
$\frac{3}{4}''$ AF high lift ring spanner for gearbox retaining bolts.
450 mm Stillson wrench.
$\frac{3}{4}''$ Drive socket bar approx 550 mm long.
A torque spanner is necessary for periodic checks on the torque limiting clutch.

2.2 *For Wolseley Swipe Jungle Buster*
Allen keys for triangle fixing screws.
$\frac{15}{16}''$ AF ring spanner.

3   *Tools for tractors*

All tractors should have a full set of the necessary spanners for the particular tractor concerned and should in addition have available on site the following items:

(i)	Adjustable spanner	(ix)	Wheel brace
(ii)	Pliers	(x)	3 tyre levers
(iii)	Screwdrivers (normal and Phillips)	(xi)	Foot pump
		(xii)	Puncture repair kit
(iv)	Block hammer	(xiii)	Air pressure gauge
(v)	Oil measure and can	(xiv)	Winch
(vi)	Funnel	(xv)	Ground anchor
(vii)	Grease gun with flexible connector	(xvi)	Drain crossing bridges
		(xvii)	Spade
(viii)	Hydraulic jack	(xviii)	Saw

Decrease in output for obstacles in weeding row

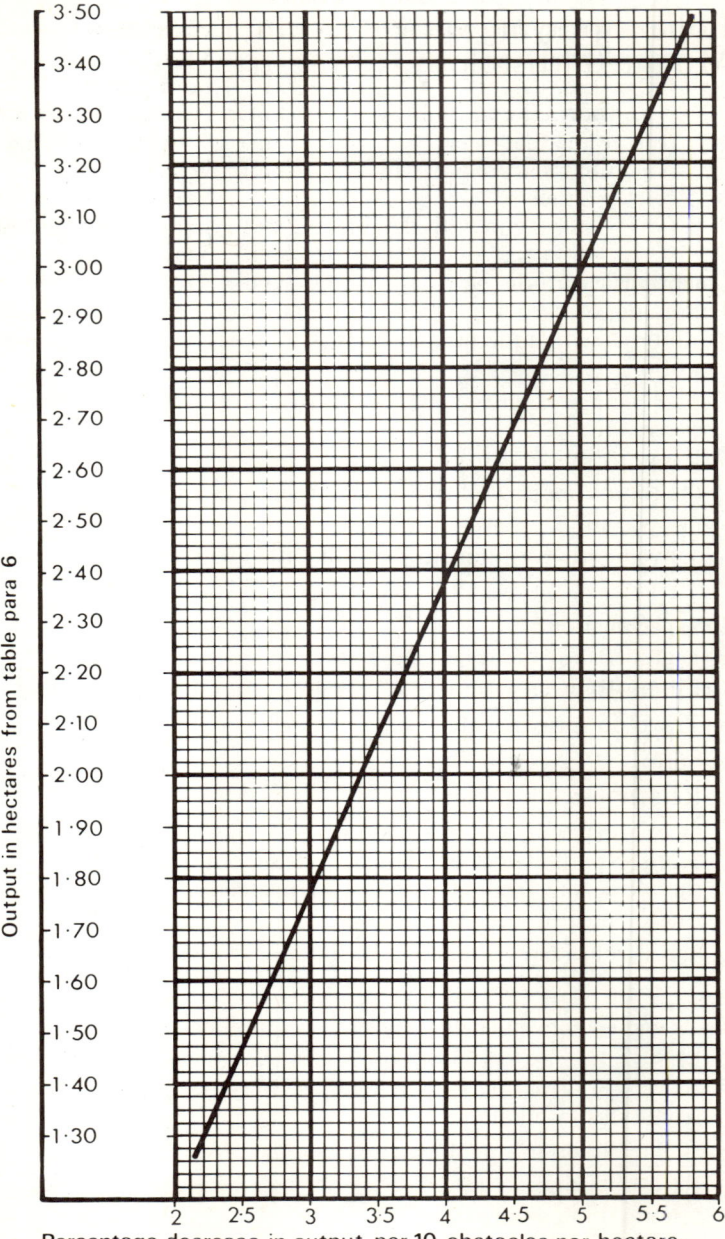

Output in hectares from table para 6

Percentage decrease in output per 10 obstacles per hectare

Foliar Spraying with a Victair Tractor-Mounted Mistblower: Forestry Model

1 Conditions

This guide refers to foliar spraying of hardwood regrowth in conifer plantations with aqueous solutions of 2, 4, 5-T under the following conditions:

a. The average height of plantation trees and woody weeds not to exceed 2 m.

b. Racks to be cut at 10 m or 15 m spacings or tractor to travel over the top of one row of trees or between two rows.

c. Drums or bowsers of diluent to be previously positioned conveniently for filling and refilling the mistblower.

2 Job specification

This guide is for the following work:

a. Fill mistblower with self-filling pump from drums or bowser.

b. Travel to spraying site.

c. Spray.

d. Return to refill.

e. Occasional 'in work' maintenance and adjust protective clothing. Note: No allowance is made for setting out drums, travelling to the work site from depot, on/off protective clothing or daily tractor/mistblower maintenance.

f. See also Forestry Commission Bulletin No 48 chapter 14.

3 Tools and equipment (including safety equipment)

a. Massey Ferguson 165 or equivalent tractor with approved cab. A MF 135 tractor or its equivalent may be used if fitted with front wheel weights.

b. Victair Standard mistblower as modified for forestry use.

c. Barrels, drums or bowsers set out at convenient intervals containing diluted herbicide.

d. Protective clothing. (See Entopath Chemical Control Supplement).

e. Washing facilities.

4 Allowances

a. Work other than that detailed in paragraphs 2a–e above is not included in the job specification and should be paid for separately.

b. No Relaxation Allowance is included as such. It is considered adequate rest can be taken during the 20 minutes refill time while the operator waits for the machine to complete the job. This period is equivalent to an allowance of 20% or more in all cases given in paragraph 6 below, except the lowest output figure of 1·64 ha/h. To raise this to 20% also would drop this output figure to 1·56 ha/h.

5 Selecting the output

a. Decide the appropriate application rate. If this falls between 100 and 150 litres/ha interpolate as necessary.

b. From the density and height of the vegetation decide the width of treatment. Normally this is 15 m but where vegetation is over 2 m and/or very dense a 10 m interval may be adopted.

c. Carry out a trial run in the compartment to determine the speed at which the tractor can be driven steadily.

d. Using these parameters read the output from the table in paragraph 6 below.

6 Output table

Tractor speed (km/h)	Output in hectares per hour			
	100 litres/ha		150 litres/ha	
	10 m strip	15 m strip	10 m strip	15 m strip
2	1·64*	2·27	1·51	2·03
3	2·27	3·03	2·02	2·60
4	2·79	3·65	2·43	3·04

*Note: see paragraph 4b.

7 Modifications
None.

Applying Atrazine using a Knapsack Sprayer

1 Conditions

a. The rows of trees must be easily visible. For spot spraying individual trees must be visible.

b. The slope must be sufficiently gentle for 3·2 kph to be maintained.

c. Some brash, litter etc may be present providing 3·2 kph can be maintained.

d. Water available in bowser or 45 gallon (about 110 litre) drums.

2 Job specification

The Output Guide applies to the following work:

a. Spray the herbicide over the row of trees maintaining the desired spraying width (usually 1 metre).

b. Walk at 3·2 kph when spraying (1 yard (0·9 m) in 1 sec).

c. Weigh or measure the required amount of powder and mix thoroughly. (See Entopath Chemical Control Supplement).

d. Maintain the pressure on which the flow was calculated eg to give a spraying time of $7\frac{1}{2}$–8 minutes.

e. Fill the sprayer to 18 litres.

f. Rinse out the sprayer at the end of the day.

g. Avoid spraying, or rinsing out, any chemical into ditches, streams etc.

3 Tools and equipment (including safety equipment)

a. Cooper Pegler, Policlair Mark II, or CP III (Forestry Model).

b. Red polijet, pressure control valve and gauge.

c. 21″ (53 cm) or 28″ (71 cm) lance.

d. Semi-rotary pump with barrel adaptor.

e. Support stand for loading or end filling drum.

f. Plastic bucket for mixing.

g. Scoop for chemical powder marked to hold desired amount (say 360 grammes) or pre-measured sachets for each sprayer-full.

h. Protective clothing. (See Entopath Chemical Control Supplement).

i. Washing facilities.

4 Allowances

a. For contingencies and work other than spraying an allowance of $17\frac{1}{2}\%$ of cyclic work has been included.

b. For personal needs and rest an allowance of 22% of total work has been included.

5 Method of selecting appropriate output
See para 6.

6 Output
The following output can be expected in a full working day of 480 minutes.

Where the rows are easily visible, walking conditions good and average distance each way to refill is:	Number of sprayers-full per day (1 sprayer-full = 18 litres)
50 metres	19–21
100 metres	17–19
150 metres	15–17

7 Modifications and variations
N.B. These modifications and variations are only to be applied when the prevailing conditions or job specification differ from those listed in paragraphs 1 and 2.

a. *Ground conditions*
Where there are impediments to movement necessitating considerable avoiding action, output is likely to be reduced by 1 or 2 sprayers-full per day.

b. *Rows difficult to follow*
Where rows are indistinct output is likely to be reduced by 1 or 2 sprayers-full per day.

c. *Steep slopes*
Where steep slopes are encountered and walking speed is reduced to 1·5kph output will fall as follows:

Where the average walking distance each way is	Number of sprayers-full per day
50 metres	13–14
100 metres	11–13
150 metres	9–11

d. *Reduced quantity carried*

The capacity of the sprayer is 18 litres. Where circumstances demand that less is carried, the effect on output is as follows:

Volume carried	Average distance walked	Number of sprayers-full per day
$\frac{3}{4}$ full (14 litres)	50 metres	23–25
	100 metres	20–22
	150 metres	18–19
$\frac{1}{2}$ full (9 litres)	50 metres	29–31
	100 metres	25–27
	150 metres	21–23

e. *Spot treatment*

If the technique is practicable, the following outputs will apply:

Average walking distance	Treating $\frac{1}{2}$ of row	Treating $\frac{3}{4}$ of row
	Number of sprayers-full	Number of sprayers-full
50 metres	$12\frac{1}{2}$–$13\frac{1}{2}$	16–17
100 metres	12 –13	14–15
150 metres	11 –12	13–14

The Application of Chlorthiamid (Prefix)

1 **Conditions**
   a. The rows of trees and, for spot treatment the individual trees, are readily visible.
   b. Slope and ground conditions such that a walking speed of 3·2 kph can be maintained.
   c. The herbicide, in 25 kg bags or kegs close to site.

2 **Job specification**
   a. Fill the hopper with one third of the contents of a 25 kg keg.
   b. Distribute the herbicide over the trees maintaining the desired width (1 metre) and desired walking speed (3·2 kph = 2 mph = 1 yd per sec).
   c. Maintain the engine revolutions at maximum allowed and fully open the on/off lever while spraying.
   d. In damp weather check that the tube below the hopper does not become partially blocked.
   e. See also FC Bulletin No 48 chapter 12.

3 **Tools and equipment** (including safety equipment)
   a. Horstine Farmery knapsack distributor with fishtail nozzle. If spot treatment is to be practised, a modified on/off handle is required.
   b. Supply of fuel, and filter funnel.
   c. A stand of some sort to facilitate lifting the distributor on to the operator's back.
   d. Plug spanner, screwdriver, carburettor spanner, adhesive tape, spare plugs.
   e. Protective clothing. (See Entopath Chemical Control Supplement).
   f. Washing facilities.

4 Allowances

a. For contingencies and work other than actually applying prefix an allowance of 15% of cyclic work has been included.

b. For personal needs and rest an allowance of 25% of total work has been included.

5 Method of calculating output

Assess the distance between refill point and midway into the area to be treated. Read off the output from para 6 below; adjust output as appropriate if conditions specified in para 7 apply.

6 Table of outputs

These outputs can be expected given the conditions specified in paragraph 1.

Average distance between refill point and midpoint of area to be treated (metres)	Output in full day	
	kgs	hectares (net)
50	80–90	1·3 –1·45
100	75–85	1·2 –1·35
150	70–80	1·15–1·3

7 Modifications and variations

N.B. These modifications and variations are only to be applied when the prevailing conditions or job specification differ from those listed in paragraphs 1 and 2.

Obstacles and short rows

a. Where there are obstacles to avoid, or short rows output will be reduced, in relation to their significance by up to 10%.

b. *Spot treatment.* Where spot treatment is practicable and 50% of each row is treated, these outputs may be anticipated.

Average distance between refill point and midpoint of area to be treated (metres)	Output in full day	
	kgs	hectares (net)
50	42–50	0·68–0·81
100	41–49	0·66–0·74
150	40–48	0·64–0·77

Spot Spraying Using Knapsack Sprayers Fitted with a Guard

1 Conditions

The times apply to spraying under the following conditions:

a. Slopes up to 15° (27%).

b. Occasional stones and rough ground.

c. Stumps and some litter from the previous crop.

d. Trees readily visible among the vegetation.

e. Adequate supplies of pre-mixed chemical solution distributed to avoid excessive walking to refill the sprayer (not more than 30 metres).

f. Where conditions differ from those shown above extra time may be required—see paragraph 7.

2 Job specification

The times apply to the following work:

a. Spraying diluted herbicide at a rate of 450 litres per treated hectare.

b. Walking to and from the spraying area to the bulk supply point.

c. Refilling knapsack sprayer from bulk container, washing out and regularly maintaining the sprayer.

d. See also FC Bulletin No 48 Chapter 9.

3 Tools and equipment (including safety equipment)

a. A Cooper Pegler CP III (Forestry Model) or Policlair Mk II knapsack sprayer fitted with pressure control valve, pressure gauge and guard. The sprayer to operate at a pressure of between $0.28 \, kg/cm^2$ and $0.42 \, kg/cm^2$ (4–6 psi).

N.B. The guard can contain either the tree (eg Politec, Policone) or the nozzle (eg Cooper Pegler internal spray shield, washing up bowl etc).

b. Either a bowser or drum(s) for the supply of pre-mixed herbicide.

47

c. Protective clothing (see Entopath Chemical Control Supplement).

d. Washing facilities.

4 Allowances

Included in the times

a. For contingencies and work other than that included in the cyclic time, eg maintenance, on/off protective clothing, washing hands etc 7% of the time spent spraying.

b. For personal needs and rest 25% of the total working time.

5 Method of using the Output Guide

Paragraph 6 gives the time per hectare against plant spacing and the type of guard being used.

Example

An area of 1·2 metres around each tree in a compartment planted at 2·1 m × 2·1 m spacing is to be sprayed with paraquat using a guarded nozzle. Conditions are as stated in paragraph 1.

Time per hectare = 398 standard minutes.

6 Time per hectare in standard minutes

(See table opposite)

Standard minutes per hectare

Tree spacing in metres	Number of trees per hectare	Using tree guard		Using guarded nozzle					
		0·9m* (42cc)	litres applied per ha	0·9m* (42cc)	litres applied per ha	1·2m* (63cc)	litres applied per ha	1·5m* (84cc)	litres applied per ha
1·4 × 1·4	5,102	742	214	557	214	693	321	767	429
1·5 × 1·5	4,444	689	187	524	187	646	280	717	373
1·6 × 1·6	3,906	639	164	491	164	601	246	670	328
1·7 × 1·7	3,460	588	145	461	145	558	218	626	291
1·8 × 1·8	3,086	540	130	430	130	517	194	583	259
1·9 × 1·9	2,770	492	116	401	116	477	175	542	233
2·0 × 2·0	2,500	445	105	373	105	437	158	501	210
2·1 × 2·1	2,268	398	95	344	95	398	143	465	191

*Diameter of the area treated around each tree (equivalent dosage per tree in brackets).

Footnote:
Where tree spacings are other than those shown above, entry to the table should be made using the appropriate number of trees per hectare and interpolating where necessary.

7 Modifications and variations to the times

N.B. These modifications and variations are only to be applied when the prevailing conditions or job specification differ from those listed in paragraphs 1 and 2.

a. *Poor ground conditions*

Where ground conditions are worse than those shown under paragraph 1, eg lop and top from previous crop not burnt, fallen boughs from ringed trees, mine-workings, unstable slopes etc
Add up to 10%
In the most difficult conditions
Add up to 20%

b. *Steep slopes*

On slopes steeper than 15° add to the times as follows:

Slope	% increase
Over 15°–19° (27%–34%)	5%
20°–24° (35%–44%)	10%
25° and over (45% and over)	15%

c. *Excess walking*

On sites where it is not possible to position the refill point within 30 metres of the spraying area
Add 0·1 standard minutes per litre sprayed for every additional 30 metres walked

d. *Allowance for moving semi-rotary pump, stand and hose*

When chemical solution is being supplied from a trailer/tank or drum by means of a semi-rotary pump and is required to be repositioned periodically
Add 5%

e. *Small trees*

Where trees are small and difficult to find under dense cover:
(i) Conifers in herbs and broadleaves in grasses
Add up to a maximum of 10%
(ii) Conifers in grasses and broadleaves in herbs
Add up to a maximum of 20%

Foliage Spraying with 50% 2, 4, 5-T in Water using Knapsack Sprayers

1 **Conditions**

The times apply to foliage spraying under the following conditions:

a. Pre- or post-planting areas.

b. All foliage to be readily accessible for spraying, plantations to be racked if necessary.

c. Slopes up to $15°$ (27%).

d. Occasional brash, litter, drains and rough ground.

e. Adequate supplies of pre-mixed chemical solution distributed to avoid excessive walking to re-fill (30 metres maximum), but see paragraph 7a.

2 **Job specification**

The times are for the following work:

a. Apply herbicide to leaf surface of unwanted weed species, avoiding planted trees as far as possible.

b. Walk to and from spray area to diluent container.

c. Refill knapsack sprayer from container, wash out and regularly maintain.

d. See also FC Bulletin No 48 chapter 9.

3 **Tools and equipment** (including safety equipment)

a. Cooper Pegler CP III (Forestry Model) or Policlair Mark II knapsack sprayer fitted with pressure control valve and pressure gauge, with pressure set at 0.70kg/cm^2 (10 psi).

b. A 53 cm (21 inch) or 71 cm (28 inch) hand lance with red polijet or politip or brass PP 78 floodjet.

c. No 1 semi-rotary pump with barrel adapter or alkathene hose.

d. Protective clothing. (See Entopath Chemical Control Supplement).

e. Washing facilities.

4 Allowances

Included in the times

a. For contingencies and work other than cyclic time, eg on/off protective clothing, washing hands etc 7% of the time spent spraying.

b. For personal needs and rest 25% of the total working time.

5 Method of using the Output Guide

Paragraph 6 gives the time to spray one litre of herbicide, to convert this to output in hectares per day divide the number of litres sprayed per day by the average sprayed volume per hectare, obtained by calibration runs.

6 Time for spraying foliage

The time per litre for applying a sprayer full of herbicide is 1·0 minute.

7 Modifications and variations to the times

N.B. These modifications and variations are only to be applied when the prevailing conditions or job specification differ from those listed in paragraphs 1 and 2.

a. Excess walking over 30 metres from chemical supply point to nearest point of spray area

Add 0·1 minutes per litre for every additional 30 metres walked.

b. *Ground conditions*

Where obstructions to walking are more than are normally present on a forest floor, eg lop and top from previous crops, fallen branches from ringed trees, mine workings, unstable slopes, poor racks etc

Add to the time per litre by up to 10%.

In the most difficult conditions, **add** up to 20%.

c. *Slope*

The time in paragraph 6 is for slopes of up to 15°. On steeper slopes increase the time as follows:

Slope	Add
Over 15°–19° (27%–34%)	5%
20°–24° (35%–44%)	10%
25°–30° (45%–58%)	15%

d. *Moving semi-rotary pump, stand and hose*

When chemical solution is being supplied from a trailer/tank or drum by means of a semi-rotary pump and barrel adapter or alkathene hose, this has to be repositioned periodically.

Add 5%.

Spraying Herbicides with Knapsack Mistblowers

1 Conditions
The times apply to mistblowing chemical solutions under the following conditions:

a. Slopes up to 15° (27%).

b. Occasional loose stones, rough ground and stumps and litter from racking.

c. In overhead cover, bramble, dense grass or coppice, rackways will be provided.

d. Adequate supplies of pre-mixed chemical solution, distributed to avoid excessive walking to refill (30 metres maximum) but see paragraph 7a.

2 Job specification
The times are for the following work:

a. Applying mist to areas of all unwanted weed species (as directed by the supervisor).

b. Filling mistblower container and fuel tank, washing out, cleaning and regular maintenance.

c. Mechanical breakdowns of less than 15 minutes.

d. See also FC Bulletin No 48 Chapter 11.

3 Tools and equipment (including safety equipment)

a. Mistblower: Stihl SG 17; Wambo S 170a; Danarm W 170; KEF Motoblo 35; Cooper Pegler CP 40; Allman L 35.

b. Fuel for mistblower, filter funnel, spare sparking plug, starter cord (where applicable), plug spanner, screwdriver, instruction book, water for cleaning mistblower and a supply of rag or cotton waste.

c. No 1 semi-rotary pump with barrel adapter or alkathene hose.

d. Protective clothing (see Entopath Chemical Control Supplement).

e. Washing facilities.

4 Allowances

Included in the times

a. For contingencies and work other than actual mistblowing eg maintenance, breakdowns of up to 15 minutes, refilling, re-fuelling, on/off protective clothing, cleaning machine and washing hands 25% of the time spent mistblowing.

b. For personal needs and rest 25% of the total working time.

5 Method of selecting the time

By means of a Calibration Run (see Bulletin No 48 p 33), determine the average rate of application and the jet size or nozzle setting to be used.

Example

Using a Wambo S 170 mistblower, calibration runs have determined that at nozzle setting $\frac{1}{4}$, the volume sprayed per hectare is 224 litres. Paragraph 6 gives the time per litre of 4·0 minutes and an output per day of 120 litres.

Output in hectares per day can be calculated by dividing the appropriate value in column 7 of paragraph 6 by the number of litres sprayed per hectare as determined by the calibration run.

eg $\dfrac{120 \text{ litres per day}}{224 \text{ litres per hectare}} = 0·54$ hectares per day.

6 Time for mistblowing

	Jet size or nozzle marking				Minutes per litre	Output in litres per day
CP 40	KEF Motoblo 35	Allman L 35	Danarm W 170 Wambo S 170a	Stihl SG 17		
–	1	2	–	–	5·3	91
2	–	–	$\frac{1}{4}$	–	4·0	120
3	–	–	–	–	3·1	155
–	–	3	–	1	2·6	185
4	2	–	–	–	2·2	218
–	–	–	–	–	2·0	240
5	–	4	$\frac{1}{2}$	$1\frac{1}{2}$	1·8	267

N.B. With the Allman L 35, Stihl SG 17, Wambo S 170a and Danarm W 170 mistblowers only, it is possible to set the output control between the values marked. The output volume can be determined by calibration runs and the times interpolated.

7 Modifications and variations to the times

N.B. These modifications and variations are only to be applied when the prevailing conditions or job specification differ from those listed in paragraphs 1 and 2.

a. *Excess walking*

For walking over 30 metres from chemical supply point to the nearest point of spray area

Add 0·1 min per litre for every additional 30 metres walked.

b. *Ground conditions*

Where obstructions to walking are more than normally present on a forest floor, eg lop and top from previous crop, fallen branches from ringed trees, mine workings, unstable slopes, poor racks etc.

Increase the time per litre by up to 10%.

In the most difficult conditions add up to 20%.

c. *Slope*

The times are for slopes up to 15° (27%) on steeper slopes.

Increase the time per litre as shown below.

Slope		Add
15°–19°	27%–34%	5%
20°–24°	35%–44%	10%
25°–30°	45%–58%	15%

d. *Allowance for using semi-rotary pump, stand and hose*

When the herbicide is supplied from a trailer/tank or drum by means of a semi-rotary pump and alkathene hose and this requires to be re-positioned periodically, the time per litre may be increased by 5%.

Cut Stump and Basal Bark Treatment with 100% 2,4,5-T in Oil Using Knapsack Sprayer or Live Reel Sprayer

Section	XV
No	G8

1 Conditions

The times apply under the following conditions:

a. All stumps, poles or coppice to be readily accessible, ie reasonable walking conditions, some bramble, brash, drains and rough ground (see paragraph 7b).

b. Bark to be dry at the time of spraying.

c. Slopes up to 25% (14°) (see paragraph 7c).

d. Required amount of chemical/oil to be available in a bowser or 45 gallon (110 litre) drums at operating positions at not more than 30 metres from the spraying area (see paragraph 7a).

e. With live reel sprayer: a two or four man team.

2 Job specification

The times are for the following work:

a. Spray all cut stumps to saturate the whole stump surface to run-off.

b. Spray all standing poles/coppice as required to saturate the bark over the full circumference of the stem from a height of 30 centimetres to ground level.

c. Spray with rose (solid stream) jet at 5 psi.

d. See also FC Bulletin No 48 Chapters 9 and 13.

Knapsack sprayer only

e. Walk from spraying site to container to refill.

f. Refill knapsack sprayer from container by semi-rotary pump.

Live reel sprayer only

g. Spray area (see a. and b. above) including hauling out main hose and handling side lines.

h. Maintain live reel sprayer.

N.B. The times do not include the following work (see para 7).

1. Prepare and set up live reel sprayer for spraying (7d).
2. Move live reel sprayer as required to next spraying area (7e).
3. Reel in and pack up on completion of spraying for the day (7f).

3 **Tools and equipment** (including safety equipment)
    a. Knapsack sprayer (18 litres capacity), Cooper Pegler, Policlair Mk II or CP III (forestry model).
or
    b. Live reel sprayer with:
       150 metre main line
       3 metre suction and return hoses
       2×37 metre side lines
       Y piece fitted with self-sealing couplings
       Tool box, tools and spares
       Operator's instruction book
The following additional equipment will be required when live reel sprayer is used by a four-man team.
       Three-way (X) junction
       2 side lines and lances
       25 and 6 metre extension hoses

    c. Pressure control valve and gauge.
    d. 21 inch (53 cm) or 28 inch (71 cm) lance.
    e. No 250 nozzle and rose jet ($5 \times \cdot 040''$ solid stream disc).
    f. No 1 semi-rotary pump and barrel adapter. (Knapsack sprayer only.)
    g. Protective clothing (see Entopath Chemical Control Supplement).
    h. Washing facilities.

Allowances included in the times

Knapsack sprayers

a. For contingencies and work other than spraying, walking and refilling, ie maintenance on/off protective clothing, washing hands etc
    15% of the cyclic time.
b. For personal needs and rest
    25% of the total working time.

Live reel sprayer

a. For contingencies and work other than spraying, walking and handling hose
    17% of the cyclic time.

b. For personal needs and rest
   17% of the total working time.

5 Method of using the Output Guide

a. Paragraph 6A deals with prep ground using either knapsack or live reel sprayers. Entry to the tables is based on an estimate of the volume of liquid applied per hectare. The time to apply 1 litre of herbicide, the time per hectare and the output in litres per hour are given.

b. Paragraph 6B deals with weeding or cleaning using either knapsack or live reel sprayers. Table layout is as outlined in 5a above.

Example

a. Prep ground.

b. Knapsack sprayer.

c. Slope 30% (17°)—other conditions as shown in paragraph 1.

d. Rate of herbicide application—400 litres per hectare.

Standard minutes per litre	= 2·11 [From paragraph 6A(i)]
Add 5% for slope	= 0·11 (From paragraph 7c)
Total time per litre	= 2·22 standard minutes
Standard minutes per hectare	= 844 [From paragraph 6A(i)]
Add 5% for slope	= 42 (From paragraph 7c)
Total time per hectare	= 886 standard minutes
Output in litres per hour	= 28·4 [From paragraph 6A(i)]
Subtract 5% for slope	= 1·4 (From paragraph 7c)
Total output	= 27·0 litres per hour

6 A Prep ground

Time to spray 1 litre of herbicide, time to spray 1 hectare with herbicide and output in litres per hour.

(i) *Knapsack sprayer*

Litres applied per hectare	Standard minutes per litre	Standard minutes per hectare	Output in litres per hour
300	2·39	717	25·1
400	2·11	844	28·4
500	1·95	975	30·8
600	1·85	1,110	32·4
700	1·79	1,253	33·5
800	1·75	1,400	34·3
900	1·71	1,539	35·1

(ii) *Live reel sprayer*

Litres applied per hectare	Standard minutes per litre	Standard minutes per hectare	Output in litres per hour
300	2·06	618	29·1
400	1·83	732	32·8
500	1·68	840	35·7
600	1·59	954	37·7
700	1·52	1,064	39·5
800	1·47	1,176	40·8
900	1·44	1,296	41·7

N.B. Interpolate where necessary.

6 B Weeding or cleaning

Time to spray 1 litre of herbicide, time to spray 1 hectare with herbicide and output in litres per hour.

(i) *Knapsack sprayer*

Litres applied per hectare	Standard minutes per litre	Standard minutes per hectare	Output in litres per hour
300	2·68	804	22·4
400	2·38	952	25·2
500	2·16	1,080	27·8
600	2·03	1,218	29·6
700	1·95	1,365	30·8
800	1·90	1,520	31·6
900	1·86	1,674	32·3

(ii) *Live reel sprayer*

Litres applied per hectare	Standard minutes per litre	Standard minutes per hectare	Output in litres per hour
300	2·34	702	25·6
400	2·06	824	29·1
500	1·87	935	32·1
600	1·76	1,056	34·1
700	1·68	1,176	35·7
800	1·63	1,304	36·8
900	1·59	1,431	37·7

N.B. Interpolate where necessary.

7 Modifications and variations to the times and outputs

N.B. These modifications and variations are only to be applied when the prevailing conditions or job specification differ from those listed in paragraphs 1 and 2.

a. *Excess walking*

On sites where it is not possible to position the refill point for knapsack sprayers within 30 metres of the spraying area

Add 0·1 minutes per litre for every additional 30 metres walked.

b. *Ground conditions*

Where ground conditions are worse than those shown in paragraph 1a and there are more obstructions to walking, eg lop and top left from previous crop, large boulders, mine-workings, fallen boughs from ringed trees etc

Add up to 10% in steps of 5%, to the times and reduce the outputs by a similar amount.

In the most difficult conditions

Add up to 20% in steps of 5% to the times and reduce the outputs by a similar amount.

c. *Slope*

When slopes are steeper than 25% (14°) increase the times or reduce the output as follows:

Slope	*% Increase in time* (or reduction in output)
25–34% (14°–19°)	5%
35–44% (20°–24°)	10%
45% and over (25° and over)	15%

d. *Prepare and set up live reel sprayer at commencement of spraying*

Allow 10 minutes per man.

e. *Move live reel sprayer*

On completion of spraying in a stint, when main line is reeled in and live reel sprayer moved normal distance of 60 metres

Allow 20 minutes per man.

Where the length of the spraying stint is over 130 metres additional time is required to walk back to live reel sprayer and haul in the main line

Allow 25 minutes per man.

f. *Pack up live reel sprayer at the end of the working period*

Allow 10 minutes per man.

Tree Injection Using a Swedish Water Pistol and a Boy Scout Axe

1 Conditions

The Output Guide applies to the application of herbicides by Swedish water pistol tree injector and Boy Scout axe under the following conditions:

a. Diluted herbicide is available in easily transportable containers, eg 5 or 20 litre (1 or 5 gallon) drums.

b. The ground conditions are classified under one of the three categories shown below:
EASY—clear floor; no restriction to movement, some light bramble.
MODERATE—some restriction to movement but not enough to warrant extensive racking. Condition between easy and difficult.
DIFFICULT—typical 'cleaning' situation. Heavy growth of coppice and bramble under light overhead shade.

c. In the densest areas, racks to be provided every 3 to 5 rows.

d. The trees to be injected should have stems individually accessible to the injector and have diameters at the injection point of between 7·5 and 20 cm.

e. The density of trees to be injected not to exceed 1,000 stems per hectare.

2 Job specification

The Output Guide applies to the following work:

a. Cuts are to be placed at 75 mm centres on birch and susceptible species and 50 mm centres on moderately resistant species such as hazel and oak. (See Entopath Chemical Control Supplement.)

b. The dose per cut to be 1 millilitre of diluted herbicide.

c. Injections to reach the outer sapwood and to be made at a convenient height on the tree (can be at waist level). See Entopath Chemical Control Supplement.

d. See also FC Bulletin No 48 Chapter 10.

3 **Tools and equipment** (including safety equipment)
   a. One Fickningspruta water pistol with the jet bored out to 2 mm (1/16th inch) diameter and equipped with a 2·5 litres ($\frac{1}{2}$ gallon) reservoir.
   b. One Boy Scout axe (such as Spear & Jackson type 3101).
   c. One belt.
   d. One sharpening stone.
   e. One filter funnel.
   f. One calibrated measuring vessel for filling the reservoir or for mixing herbicide and diluent.
   g. Protective clothing (see Entopath Chemical Control Supplement).

4 **Allowances included in the Output Guide**
   a. For contingencies and work other than actually injecting, eg sharpening axe etc, 8% of the injecting time.
   b. For personal needs and rest, 16% of the total working time.

5 **Method of using the Output Guide**
   a. Estimate the number of trees to be treated per hectare and assess the average girth in millimetres per tree at the injection point.
   b. Divide the average girth by either 75 (for birch and all susceptible species) or by 50 (for oak and all resistant species) to give the number of cuts per tree.
   c. Assess the ground conditions under the scale shown in paragraph 1b.
   d. Entry to the tables is gained by using the number of trees to be treated per hectare, the average number of cuts to be made per tree and the ground conditions.
   Example
   Birch on easy ground conditions.
   a. Number of trees per hectare—600.
   b. Average girth—300 millimetres.
   c. Number of cuts per tree $= \dfrac{300}{75} = 4$.

From paragraph 6A the time per litre is 111 standard minutes.
or
From paragraph 6B the time per tree is 0·44 standard minutes.

Where unwanted trees occur in mixture the following method of assessing the time may be used:

Moderate ground conditions	Birch	Oak
a. Number of trees per hectare	600	400
b. Average girth	300 mm	400 mm
Therefore total girth per hectare	180,000 mm	160,000 mm
c. Divide by factor for number of cuts	75	50
Therefore number of cuts per hectare	2,400	3,200

d. Total number of cuts for both species   5,600

e. Total number of trees for both species   1,000

f. Average number of cuts per tree equals 5·6.

From paragraph 6A the time per litre is 105 standard minutes.

or

From paragraph 6B the time per tree is 0·59 standard minutes.

6A Time in standard minutes to apply 1 litre of herbicide

Average no of cuts per tree	Easy conditions			Moderate conditions			Difficult conditions		
	1,000 trees per ha	600 trees per ha	300 trees per ha	1,000 trees per ha	600 trees per ha	300 trees per ha	1,000 trees per ha	600 trees per ha	300 trees per ha
3	111	119	132	120	132	150	147	165	200
4	107	111	119	115	122	135	132	145	175
5	102	105	112	108	114	123	123	134	154
6	97	100	107	103	108	116	115	124	141
7	93	96	102	98	93	109	109	115	132
8	90	93	97	94	99	104	104	110	124
9	87	89	94	90	94	100	100	105	116
10	83	86	89	87	90	95	95	100	110
11	80	82	85	84	87	91	91	95	104
12	78	79	82	81	83	87	87	91	99
13	75	76	78	77	79	83	83	87	94
14	72	73	75	74	75	79	79	83	90
15	68	69	73	71	73	76	76	79	86
16	65	66	71	69	71	73	73	75	82

Interpolate where necessary.

6B Time in standard minutes to treat one tree

Average no of cuts per tree	Easy conditions			Moderate conditions			Difficult conditions		
	1,000 trees per ha	600 trees per ha	300 trees per ha	1,000 trees per ha	600 trees per ha	300 trees per ha	1,000 trees per ha	600 trees per ha	300 trees per ha
3	0·33	0·36	0·40	0·36	0·40	0·45	0·44	0·50	0·60
4	0·43	0·44	0·48	0·46	0·49	0·54	0·53	0·58	0·70
5	0·51	0·52	0·56	0·54	0·57	0·61	0·61	0·67	0·77
6	0·58	0·60	0·64	0·62	0·65	0·70	0·69	0·74	0·85
7	0·65	0·67	0·71	0·69	0·72	0·76	0·76	0·80	0·92
8	0·72	0·74	0·78	0·75	0·79	0·83	0·83	0·88	0·99
9	0·78	0·80	0·85	0·81	0·85	0·90	0·90	0·94	1·04
10	0·83	0·86	0·89	0·87	0·90	0·95	0·95	1·00	1·10
11	0·88	0·90	0·93	0·92	0·96	1·00	1·00	1·04	1·14
12	0·94	0·95	0·98	0·97	1·00	1·04	1·04	1·09	1·19
13	0·98	0·99	1·01	1·00	1·03	1·08	1·08	1·13	1·22
14	1·00	1·02	1·05	1·04	1·05	1·11	1·11	1·16	1·26
15	1·02	1·03	1·09	1·06	1·09	1·14	1·14	1·19	1·29
16	1·04	1·06	1·14	1·10	1·14	1·17	1·17	1·20	1·31

Interpolate where necessary.

7 Modifications and variations to the times

N.B. These modifications and variations are only to be applied when the prevailing conditions or job specification differ from those listed in paragraphs 1 and 2.

a. If regrowth from the unwanted stems has to be cut before injection can take place extra time will be required. Consult Work Study Branch.

Weeding and Cleaning
by Hand Tools or Portable Brushcutter

1 Conditions

Considerable confusion can arise between the terms weeding and cleaning. As a general guide when using this Output Guide, weeding is defined as the cutting of annual growth before the establishment of the crop, whilst cleaning is the cutting of several year's growth after establishment. However full use should be made of the weed description tables given in the Appendix when determining which time table to use.

The times apply to weeding and cleaning under the following conditions:

a. Obstructions to walking as normally present in a weeding or cleaning situation, ie overhead cover trees, stumps, lop and top or litter from previous stand, drains, occasional rough ground and loose stones. (See paragraph 7a for more difficult conditions.)

b. Slopes of up to 25% (14°) (see paragraph 7b for steeper slopes).

Weeding only

c. Few trees visible above the surrounding vegetation but fairly easily found (see paragraph 7c for easier or more difficult visibility).

d. Removal of one season's weed growth carried out at any time of the year (see paragraph 7f for more than one season's growth).

Cleaning only

e. Trees generally above the level of the grass and herbaceous weeds but suffering competition from climbers and woody weeds. (See paragraph 7c for more difficult visibility.)

f. Weeds to be cut in late winter or early spring when the grass and herbs are dead and there are no leaves on the woody growth (see paragraph 7i when cleaning is carried out during the growing season). Removal of several years growth (typically 3–5 years).

2 Job specification

The times apply to the following work:

a. Cut vegetation to be kept clear of drains, footpaths, racks and rides.

Weeding only

b. Cut a swath 1 metre wide around the trees leaving the stubble 15–25 cms high between plants and 8–15 cms high around plants. Inter row vegetation to be cut only if it is high enough, or is likely to become high enough to fall on the trees or to adversely influence their growth.

Cleaning only

c. All climbers and woody weeds to be cut and laid between rows.

d. See also FC Bulletin No 48 Chapters 1–3 and 5.

3 Tools and equipment (including safety equipment)

a. Protective clothing viz:
   For hand work
   Industrial gloves.
   Boots with safety toecap.

   When using a clearing saw
   Safety helmet BS 2826, 5240 or 2095.
   Ear defenders.
   Eye protection.
   Industrial gloves.
   Boots with safety toecap.

Weeding only

b. By hand:
   Where vegetation is predominantly groups A or B, a reap hook or Dutch weeding scythe. For heavier vegetation, a light brushing hook or reap hook with a long handle or bean hook.

c. By portable brushcutter:
   Husqvarna Jonsereds or other machine fitted with acceptable vibration dampening.
   Two spare blades.
   Fuel cans.
   Spare sparking plug.
   Filler funnel with filter.
   Feeler gauge.
   Box in which to store machine.
   Maintenance schedule.

Cleaning only

d. By hand:
   Where vegetation is a mixture of climbers and woody weeds

a light brushing hook. Where mainly thick woody weeds, a heavy brushing hook or S-hook.

e. By portable brushcutter:
   As b above.

4 Allowances

The following allowances are included in the times:

a. For contingencies and work other than the actual cutting time, eg tool maintenance, general preparation, walking between working sites, unavoidable delays and in the case of portable brushcutters breakdowns of up to 30 minutes accumulated duration per day. Weeding and cleaning by hand tools 10% of the time spent cutting. Weeding and cleaning by brushcutter 21% of the time spent cutting.

b. For personal needs and rest 15–20% of the total working time.

5 Method of using the Output Guide

a. Define:
   1 The major weed groups present:
      A. Bracken, herbaceous plants and soft grasses.
      B. Coarse grasses and rush.
      C. Climbers, mostly bramble.
      D. Woody weeds (Coppice and seedlings).
   2 The grade of difficulty of each group (see appendix).
   3 The cover of each group as a percentage of the area to be cut, estimated in steps of 5%.

N.B. The area may include a percentage of ground where weed growth is not competing with the crop and does not therefore require cutting. A time must however be allocated to these areas where it is impracticable to avoid walking them when cutting (see paragraph 6).

b. Convert the indices of quantity and difficulty for each weed group into a time to cut one hectare from paragraph 6 as follows:
   (i) The tables give the times per hectare for cutting the four major weed groups according to the grade of difficulty and row spacing.
   (ii) Multiply these times by the appropriate cover percentage.
   (iii) Add together the times for each of the weed groups to find the total cutting time.

Example
Weeding with hand tools.
Conditions as specified in paragraph 1.
Job specification as paragraph 2.
2·1 m × 2·1 m plant spacing.

Major weed group	Grade	Time per hectare for 100% cover (from 6A)	% cover	Time per hectare for given % $\dfrac{\text{Col 3} \times 4}{100}$
Herbaceous plants	2	1,000	50	500
Coarse grasses	1	900	10	90
Climbers	2	1,950	5	98
Woody weeds	3	3,350	20	670
Areas not requiring cutting	–	150	15	23
Total			100	1,381
				say 1,380 standard minutes

6A Weeding by hand tools—Time per hectare in standard minutes

Weed group	Grade	Plant spacing (metres)							
		2·1 × 2·1	2·0 × 2·0	1·9 × 1·9	1·8 × 1·8	1·7 × 1·7	1·6 × 1·6	1·5 × 1·5	1·4 × 1·4
A. *Herbs, bracken and most soft grasses* Eg foxglove, willow herb, meadow sweet, sorrel and thistle and other wild flowers. *Grasses* Eg agrostis, holcus, cocksfoot, non tussocky molinia etc.	1	560	590	630	670	710	750	800	850
	2	1,000	1,040	1,090	1,150	1,210	1,290	1,380	1,490
	3	1,630	1,710	1,800	1,900	2,020	2,150	2,290	2,440
B. *Coarse grasses and rush* Eg deschampsia caespitosa, calamagrostis, tussocky molinia etc.	1	900	950	1,010	1,070	1,130	1,200	1,270	1,340
	2	1,330	1,390	1,470	1,560	1,660	1,770	1,900	2,040
	3	2,070	2,160	2,270	2,400	2,550	2,710	2,890	3,090
C. *Climbers* Principally bramble, briar, also honeysuckle, clematis.	1	1,190	1,240	1,300	1,380	1,460	1,560	1,660	1,780
	2	1,950	2,050	2,160	2,290	2,430	2,590	2,760	2,950
	3	2,490	2,530	2,580	2,650	2,740	2,850	2,970	3,100
D. *Woody weeds* Coppice and seedlings, gorse and broom included.	1	1,410	1,470	1,540	1,630	1,730	1,830	1,950	2,090
	2	2,490	2,530	2,580	2,650	2,740	2,850	2,970	3,110
	3	3,350	3,390	3,480	3,620	3,800	4,030	4,320	4,650
Areas not requiring any cutting (See 5a–3)		150	155	160	170	180	190	200	210

N.B. Interpolate between grades where necessary.

6B Weeding by portable brushcutter—Time per hectare in standard minutes

Weed group	Grade	Plant spacing (metres)							
		2·1 × 2·1	2·0 × 2·0	1·9 × 1·9	1·8 × 1·8	1·7 × 1·7	1·6 × 1·6	1·5 × 1·5	1·4 × 1·4
A. *Herbs, bracken and most soft grasses*									
Eg foxglove, willow herb, meadow sweet, sorrel and thistle and other wild flowers.	1	720	770	820	870	920	970	1,030	1,080
	2	840	910	980	1,050	1,120	1,190	1,250	1,320
	3	1,040	1,120	1,200	1,280	1,350	1,430	1,500	1,570
Grasses Eg agrostis, holcus, cocksfoot, non tussocky molinia etc.									
B. *Coarse grasses and rush*									
Eg deschampsia caespitosa, calamagrostis, tussocky molinia etc.	1	620	720	820	910	1,000	1,100	1,180	1,270
	2	1,010	1,100	1,180	1,260	1,340	1,420	1,490	1,550
	3	1,210	1,310	1,400	1,500	1,600	1,690	1,780	1,870
C. *Climbers*									
Principally bramble, briar, also honeysuckle, clematis.	1	960	1,030	1,100	1,170	1,250	1,310	1,380	1,450
	2	1,240	1,330	1,420	1,510	1,610	1,710	1,810	1,910
	3	1,530	1,650	1,770	1,890	2,020	2,140	2,260	2,380
D. *Woody weeds*									
Coppice and seedlings, gorse and broom included.	1	1,040	1,120	1,200	1,280	1,350	1,430	1,500	1,570
	2	1,530	1,650	1,770	1,890	2,020	2,140	2,260	2,380
	3	2,030	2,180	2,340	2,490	2,650	2,800	2,960	3,120
Areas not requiring any cutting (See 5a–3)		150	155	160	170	180	190	200	210

N.B. Interpolate between grades where necessary.

6C Cleaning by hand tools—Time per hectare in standard minutes

Weed group	Grade	Plant spacing (metres)							
		2·1 × 2·1	2·0 × 2·0	1·9 × 1·9	1·8 × 1·8	1·7 × 1·7	1·6 × 1·6	1·5 × 1·5	1·4 × 1·4
C. *Climbers* Principally bramble, briar also honeysuckle and clematis.	4	2,620	2,710	2,810	2,900	3,000	3,100	3,210	3,310
	5	3,460	3,590	3,720	3,850	3,980	4,110	4,250	4,380
	6	4,770	4,970	5,170	5,370	5,560	5,760	5,950	6,130
D. *Woody weeds* Coppice and seedlings, gorse and broom included.	4	4,500	4,680	4,860	5,040	5,220	5,400	5,580	5,760
	5	6,570	6,840	7,110	7,380	7,640	7,910	8,170	8,420
Areas not requiring any cutting (See 5a–3)		150	155	160	170	180	190	200	210

N.B. Interpolate between grades where necessary.

6D Cleaning by portable brushcutter—Time per hectare in standard minutes

Weed group	Grade	Plant spacing (metres)							
		2·1 × 2·1	2·0 × 2·0	1·9 × 1·9	1·8 × 1·8	1·7 × 1·7	1·6 × 1·6	1·5 × 1·5	1·4 × 1·4
C. *Climbers* Principally bramble, briar also honeysuckle and clematis.	4	1,940	2,010	2,070	2,130	2,200	2,270	2,340	2,410
	5	2,420	2,550	2,670	2,780	2,890	2,990	3,090	3,180
	6	3,360	3,530	3,700	3,850	4,000	4,140	4,270	4,400
D. *Woody weeds* Coppice and seedlings, gorse and broom included.	4	3,280	3,410	3,540	3,670	3,800	3,940	4,070	4,210
	5	4,820	5,000	5,190	5,370	5,550	5,730	5,920	6,100
Areas not requiring any cutting (See 5a–3)		150	155	160	170	180	190	200	210

N.B. Interpolate between grades where necessary.

7 Modifications and variations to the times

N.B. These modifications and variations are only to be applied when the prevailing conditions or job specification differ from those listed in paragraphs 1 and 2.

a. *Below average ground conditions*

Where the obstructions to walking are more than normally present in a weeding or cleaning situation, eg excessive lop and top from a previous crop, fallen boughs from ringed trees, mine workings, unstable slopes, ploughed sites where furrows are choked with vegetation etc

Add up to 20% in steps of 5%.

b. *Steep slopes*

When slopes are steeper than 25%

Increase the times as follows:

Slope		% Increase in time
over 25–35%	(14°–19°)	5%
over 35–45%	(20°–24°)	10%
over 45%	(over 25°)	15%

c. *Visibility of trees*

Modify the times as follows:

Degree of difficulty in finding trees	Description	Modification Weeding	Cleaning
Easy	Most leaders visible above the vegetation; remainder of trees can be seen readily when the vegetation is parted.	Subtract 10%	No modification
Moderate	Few trees visible above the surrounding vegetation, but fairly easily found.	No modification	Add 10%
Difficult	Trees well below the level of the vegetation; much searching required. Usually associated with small or poorly furnished trees.	Add 10%	Add 20%
Very difficult	Trees well below the level of the vegetation; much searching required. May have to cut over vegetation to next large tree then retrace steps to find intermediate trees. (Usually associated with very small trees in thick grass.)	Add 20%	Not applicable

Note: Treat missing trees as reducing visibility.

d. *Variation to job specification*
Where the job specification requires more or less vegetation to be cut than the specification given in paragraph 2b or c

Increase or **Reduce** the times by up to 15% in steps of 5%.

e. *Ploughed ground*
On sites where weed growth is suppressed in the furrows and weeding is done by walking along the ridge

Reduce the times by up to 20% in steps of 5%.

On sites where weed growth is vigorous and widespread and where unstable ridges make it necessary to work from the furrow

Increase the times by up to 10% in steps of 5%.

f. *More than one season's growth in a weeding situation*
On sites where weeding has not been carried out for more than one season

Increase the times for weeding with hand tools by 5% or 10% per year's extra growth for coarse grasses and rush.

Increase the times in steps of 5% up to 25% per year's extra growth for climbers and woody weeds.

g. *Closed canopy (cleaning situation)*
Where the canopy or most branches have interlaced making working more difficult

Increase the times by 5% or 10%.

h. *Cleaning carried out during the growing season*
When it is necessary to carry out cleaning during the growing season when the woody growth is in leaf and grasses and herbs may also need to be cut

Increase the times by up to 10% in steps of 5%.

i. *Exceptionally heavy grass and herbaceous growth*
On sites where the full summer's weed growth is exceptionally heavy, eg dense bracken more than 2 metres high where it is necessary to pile the bracken deliberately between the rows

Increase the times for Grade 3 by 5% steps up to 20% for the extreme cases.

j. *Underplanted larch plantations*
Where there is a rank growth of *Deschampsia flexuosa, Anthoxanthum, Festuca* etc:

Increase the times for tough grasses by 5% steps up to 20%.

Guide for assessing the difficulty of cutting weed growth

For weeding, three grades of difficulty are recognised in each of the four major weed groups.

For cleaning a further three grades are recognised for climbers and two for woody weeds.

The assessment comprises two factors: the density of the weed stems and the growth characteristics ie height, thickness and toughness. The grade of difficulty is the combination of these two factors. If necessary foresters should interpolate between grades.

Example of weed assessment:

a. An area to be weeded for the second year has a growth of bramble. The bramble plants are about 1·8 m apart and the runners are thin and trailing along the ground. The area should clearly be assigned to Grade 1, light climbers.

b. The same area is to be weeded for the fourth year. The bramble plants are growing closely, only about 0·6 m apart; the previous weeding was done late in the summer and the runners are numerous but thin and short. The bramble shows some characteristics of heavy growth and some of light growth. The correct assessment would be moderate.

The characteristics described in the following tables are intended as a guide only and reasonable tolerance should be allowed in the measurements given, in particular height, which can vary considerably without seriously altering the difficulty. Generally it is the density and thickness of stems which determines the cutting times.

Note: Stem density should only be assessed in pure patches of the vegetation in question.

Grade	Height	Thickness	Density of stems	General description of factors indicating grade
A. Bracken and other ferns				
1. Light	0·6–0·8 m	5 mm	Possible to walk without crushing bracken stems.	Bracken canopy not dense enough to conceal bare ground or grass growing around the fronds. Typical of bracken sites in spring. On sites where bracken growth is light or moderate throughout the season other weeds found are typically soft grasses and herbaceous plants.
2. Moderate	0·6–1·2 m	Mostly 5–12 mm	About a boot width apart.	Bracken canopy mostly conceals bare ground, short grass and small trees. Typical of bracken sites in early summer.
3. Heavy	1·0–1·5 m	5–12 mm	Mostly 5–8 cm	Bracken canopy completely conceals bare ground and small trees. Typical growth of late summer and of sites where bracken grows well and has suppressed most other weeds. Cut fronds tend to fall back on the worker and require keeping in the rows with more deliberation than in 1 and 2.

Grade	Height	Thickness	Density of stems	General description of factors indicating grade

A. Herbaceous plants. Willow herb, foxglove, meadow sweet, spurge and other wild flowers; thistle, butterbur; soft grasses, eg agrostis, holcus, cocksfoot.

Grade	Height	Thickness	Density of stems	General description of factors indicating grade
1. Light	0·6 m		Scattered stems	Growth not dense enough to conceal ground and small trees. Typical spring growth on moist sites. Generally a site growing mostly herbs classified as light would not require weeding. Grade 1 is usually given where herbs occur in mixture with tougher weeds such as coppice and bramble.
2. Moderate	0·6–1·1 m		Up to 0·1 per sq m	Weed growth mostly dense enough to conceal ground and small trees. Flowering plants in flower. Typical summer growth on moist sites.
3. Heavy	0·9 m		More than 0·1 per sq m	Ground and small trees concealed. Flowering plants in flower or seed. Dense lower storey of soft grasses which requires cutting.

B. Coarse grasses and rush. Eg Deschampsia caespitosa.

Grade	Height	Thickness	Density of stems	General description of factors indicating grade
1. Light	1 m		Scattered stems	Thin covering of grass stalks. Would probably not require cutting if it were the only weed.
2. Moderate	Up to 1 m		Small clumps	Fairly dense covering deschampsia forming occasional awkward clumps.
3. Heavy	1 m and over		Large and frequent clumps	Dense covering. Tough and difficult to cut whether wet or dry.

N.B. Soft grasses can become tough or wiry in the autumn and thought of as coarse grass.

C. Climbers. Bramble, rosebriar, raspberry cane, honeysuckle. One year's growth.

Grade	Height	Thickness	Density of stems	General description of factors indicating grade
Weeding only				
Bramble				
1. Light	Up to 0·6m	Up to 5mm	Plants mostly more than 1·2m apart	Typical bramble colonisation in first year or two after planting eg trailing along ground occasional runners arching and beginning to interlace. Not bushy and with few or dead stems from previous year's growth.
2. Moderate	0·3–0·6m with longer runners	Mostly 5–10mm	Plants mostly 1–1·2m apart	Mostly bushy. Runners from different plants frequently interlacing. Climbs high into trees or coppice when opportunities occur.
3. Heavy	Mostly over 0·6m with long runners	Up to 12mm	Plants mostly 0·6–0·9m apart	Dense bushes, difficult to walk through. Many tough, dead stems from previous year's growth. Requires repeated strokes with the hook making rhythmic cutting difficult. Long handled hook required.

C—continued

Cleaning only—several years' growth

Grade	Height	Thickness	Density of stems	General description of factors indicating grade
4. Light	1–1·2 m with occasional large runners	Up to 12 mm	Plants mostly 0·6–0·9 m apart	Typical of sites not weeded for several years. Long runners entwined with trees and coppice.
5. Moderate	1–1·2 m with runners up to 3 m long	Up to 20 mm	Plants mostly 0·3–0·6 m apart	Bramble sometimes in waist high bushes. Honeysuckle forming occasional dense mats and climbing trees.
6. Heavy	1·2–1·5 m with runners often over 30 m long	Mostly 20 mm	Mostly under 0·3 m apart	Often dense bushes up to shoulder height. Very difficult to push through before cutting. Long runners entwined amongst coppice stems making it difficult to cut freely. Multiple honeysuckle stems invading up to 50% of the trees.

N.B. Raspberry cane forms dense patches. It is softer than bramble, does not bush so densely and rarely justifies a grading of 3 or more.

D. One year's growth. Coppice regrowth and seedlings. Thickness classification depends on the height at which growth is to be cut.

Grade	Height	Thickness	Density of stems	*General description of factors indicating grade*
Weeding only				
1. Light	Varies considerably up to 1 metre	Mostly up to 5 mm	Stumps up to 4 m or more apart along the rows. Up to a score of shoots per stump seedlings scattered.	Early season's growth. Soft, green and fairly easy to cut with reap hook. Alternatively late season's growth, but with coppice stems very thin.
2. Moderate	Up to 1·2 metres	Mostly 5–20 mm	Stumps 2·5–4 m apart along the rows 20–30 shoots per stump.	Usually late season's growth. Tough, often requires repeated blows when using a reap hook.
3. Heavy	1·2 metres and above	Mostly more than 10 mm	Stumps frequent. More than 30 shoots on many stumps.	Late season's growth. Coppice requires repeated blows. Long handled reap hook or light brushing hook preferred.

D—*continued*

Cleaning only— several years' growth

Grade	Height	Thickness	Density of steps	General description of factors indicating grade
4. Very heavy	Up to 3 metres but mostly in the range 1·2–2·5m	Up to 20mm	Stumps frequent. 30–40 shoots on most stumps.	Many stems cut with one stroke using a heavy brushing hook or S-hook. On coppice stools several stems cut with one stroke.
5.	Up to 4·5 metres but mostly in the range 2–4 metres	20–40mm	Stumps very frequent. More than 40 shoots on many stumps.	Including a small proportion of stems up to 5·0cm diameter. Most stems require more than one cutting stroke.

N.B. *Woody weeds variations in toughness*
Some species are much tougher and more difficult to cut than others, eg oak, alder, ash, hazel, broom and gorse and where these comprise the bulk of woody weed growth they may require a higher grading than the more common mixture where softer species such as willow, sycamore, birch and elder are included.

Ultra Low Volume Application of Herbicides

1 Conditions

The times apply to spraying under the following conditions:

a. Pre and post planting areas.

b. Slopes up to 25% (14°) (see para 7c).

c. An adequate supply of herbicide on the spraying site (see para 3b(i) or (ii)).

d. Vegetation and crop heights up to a maximum suitable for ULV application. See FC Leaflet No 62.

e. Ground and vegetation conditions such that the operator can proceed at the required walking pace for the particular application and chemical flow rate concerned (see para 7c).

f. Mean row lengths between 150 and 250 m (see para 7b).

2 Job specification

The times are for the following work:

a. *Silvapron T and Asulox*
Applying herbicide with a Micron ULVA or ULVA 8 at the recommended application rate in 5 m wide swaths (see para 7a).

b. *Silvapron D*
Applying herbicide with a Micron ULVA or ULVA 8 at either 10 litres/ha to each inter-row (ie 2·1 m wide swaths) in a post planting or at 15 litres/ha in 5 m wide swaths in a pre planting situation (see para 7a).

c. *Either:*
    (i) refilling 1 litre application bottle from 5 litre containers;
    (ii) setting out sufficient 5 litre refill containers and walking to attach nearest container to knapsack when required.
    Or
    (iii) fitting new 1 litre bottles to ULVA;
    (iv) setting out sufficient 1 litre refill bottles and walking to replace empty bottle in pouch with nearest full bottle when required.

d. Necessary cleaning of sprayer and ancillary equipment and day to day maintenance (see para 4 N.B.).

N.B. See also FC Leaflet No 62 for full details of method.

3 **Tools and equipment** (including safety equipment)

a. Micron ULVA or ULVA 8 together with batteries and nozzles as required.

N.B. The nozzle giving the fastest flow rate appropriate to the walking conditions in any given situation should be used (see para 7c).

b. *Either:*

    (i) knapsack and sufficient spare 5 litre containers for 1 day's spraying per man.

    Or

    (ii) sufficient 1 litre bottles together with carriers (6 bottles per carrier) for 1 day's spraying per man;

    (iii) locally made pouch to carry spare 1 litre bottle when knapsack and refill containers are not used.

c. 1 litre plastic graduated jug for calibration.

d. Funnel with gauze filter.

e. Disc speed meter, wind speed meter and smoke pellets.

f. Tool kit for maintenance of equipment (see FC Leaflet No 62 Appendix).

g. Protective clothing as specified in FC Leaflet 62 or Entopath Chemical Control Supplement.

h. Washing facilities and barrier cream.

4 **Allowances**

a. For contingencies and other work directly involved during spraying periods, eg inspect and consider, maintenance of sprayer, on/off part of protective clothing etc, varying from 5% with a nozzle flow rate of 0·50cc/sec to 23% with a nozzle flow rate of 3·5cc/sec.

b. For personal needs and rest 12% on other work and from 18–21% on cyclic work varying with nozzle flow rate.

N.B. No allowance is included for time involved with recharging batteries when required, nor for washing ULV suit. Repairs and maintenance in excess of 15 minutes per day are also not included.

5 **Method of using the Output Guide**

a. Para 6A gives the standard times per litre and per hectare for ULV spraying at various application and nozzle flow rates using the knapsack refill system.

b. Para 6B as for 6A but using 1 litre refill bottles.

c. In addition to the other work directly involved during spraying, there is additional other work, most of which is necessary before and/or after each spraying period, eg unload/load equipment, to and from camp, on/off protective clothing, calibration, check wind and disc speeds etc. For a period of from half a man day's work to a full man day, 60 SMs should be allowed and for less than half a man day, 30 SMs allowed (per man).

d. To obtain output in litres or hectares, deduct either 60 or 30 minutes from the total time available for spraying (see para 6c) and divide by the appropriate number of SMs;

eg with a nozzle flow rate of 3cc/sec, application rate of 7 litres/ha and the knapsack refill system, from table 6A, output for an 8 hour day

$$= \frac{(480-60)}{13\cdot7} = 30\cdot7 \text{ litres}$$

$$\text{or } \frac{(480-60)}{96} = 4\cdot4\,\text{ha}$$

N.B. Nozzle flow rates vary according to temperature (see FC Leaflet No 62).

6 Times for spraying

A. Knapsack refill system

Nozzle flow rate cc/sec	3·50	3·25	3·00	2·75	2·50	2·25	2·00	1·75	1·50
SMs/Litre	12·7	13·2	13·7	14·3	15·1	16·0	17·2	18·6	20·6
SMs/Ha @ 5L/Ha	64	66	69	72	76	80	86	93	103
7L/Ha	89	92	96	100	106	112	120	130	144
7.5L/Ha	95	99	103	107	113	120	129	140	155
10L/Ha	127	132	137	143	151	160	172	186	206
15L/Ha	191	198	206	215	227	240	258	279	309

B. 1 litre bottle refill system

Nozzle flow rate cc/sec	3·50	3·25	3·00	2·75	2·50	2·25	2·00	1·75	1·50	1·25	1·00	0·75	0·50
SMs/Litre	10·0	10·4	11·0	11·6	12·3	13·2	14·3	15·8	17·7	20·4	24·5	31·3	44·8
SMs/Ha @ 5L/Ha	50	52	55	58	62	66	72	79	89	102	123	157	224
7L/Ha	70	73	77	81	86	92	100	111	124	143	172	219	314
7.5L/Ha	75	78	83	87	92	99	107	119	133	153	184	235	336
10L/Ha	100	104	110	116	123	132	143	158	177	204	245	313	448
15L/Ha	150	156	165	174	185	198	215	237	266	306	368	470	672

c. N.B. Before dividing by the appropriate number of SMs, deduct the following number of minutes from the total time available for spraying per day (ie from the actual time at the spraying site).

$$\left. \begin{array}{l} \text{For from } \tfrac{1}{2}\text{–1 man day deduct 60 minutes} \\ \text{For less than } \tfrac{1}{2} \text{ man day deduct 30 minutes} \end{array} \right\} \text{ per man}$$

7 Modifications and variations to the times

N.B. These modifications and variations are only to be applied when the prevailing conditions or job specification differ from those listed in paragraphs 1 and 2.

a. *Variations in swath spacing*
  (i) Where swath spacing is increased from 5m to 6m or from 2·1m to 2·4m deduct 5% from the times.
  (ii) Where swath spacing is decreased from 5m to 4m or from 2·1m to 1·8m add 5% to the times.

b. *Variations in row length*
  Where mean row length is less than 150m add to the standard times as shown in the following table:

Application rate (Litres/Ha)	SMs/Litre to add 50m	SMs/Litre to add 100m	SMs/Ha to add 50m	SMs/Ha to add 100m
5·0	2·6	0·9		
7·0	1·9	0·6		
7·5	1·8	0·6	13·5	4·5
10·0	1·3	0·4		
15·0	0·9	0·3		

N.B. Row lengths in excess of 250m do not make a significant difference to the times.

c. *Poor ground conditions and slopes over 25% (14°)*
  Where ground conditions and/or slopes of over 25% (14°) make walking difficult with a given nozzle, but do not warrant changing to a nozzle with a slower flow rate, add up to 20% in steps of 5%.

N.B. Add this percentage to the standard times as modified by para 7b if applicable.

Tractor-Mounted Controlled Droplet Applicator

1 Conditions

The Output Guide applies to spraying flowable atrazine (Weedex A500L) under the following conditions:

a. Slopes of up to 5° (9%).

b. Bare, freshly cultivated land (ie no stumps) or destumped areas with stumps piled in windrows or clearfell areas not destumped.

c. No more than 10 stumps, standing trees or other obstacles per hectare causing delay or detour (see paragraph 7a).

d. Adequate space at the end of the rows for the tractor to turn without manoeuvring.

e. Trees planted at row spacings of between 2·2m and 2·6m and in reasonably parallel rows.

2 Job specification

The Output Guide applies to the following work (see Appendix for working method):

a. Fill, calibrate, adjust and maintain the sprayer.

b. Spray two rows at a time. Travel at a constant speed of 60m per minute. Apply liquid at the rate of 3cc per second per head.

c. Turn at the end of each pair of rows into the next pair.

d. Check that each spray head is working before commencing next pair of rows (prep spray).

3 Tools and equipment

a. Tractor (Massey Ferguson 165 or similar) and correctly mounted controlled droplet applicator.

b. Tractor fuel, oil, etc.

c. Adequate supply of atrazine.

d. At least 45 litres of clean water in two containers plus an additional container of say 10 litres, for dilution and washing out.

e. A measuring jar, graduated in 10cc steps, capable of holding 0·5 litres.

f. A watch with a sweep second hand.

g. Tools, including small crosshead (Phillips) and standard screwdrivers, pliers, wire strippers, adjustable spanner.

h. Spare spray heads, nozzles, fuses (5 amp).

i. Protective clothing as specified by the safety officer, viz:
   Ear defenders (except when tractor is fitted with an approved 'Q' cab)
   Safety boots
   *Waterproof trousers or leggings
   *Waterproof jacket
   *Waterproof gloves (not rubber)
   *Face shield
   *Particle mask

   *These items are necessary when handling atrazine and when calibrating and maintaining the controlled droplet applicator. The use of gloves is also necessary when handling tractor fuel oil.

j. Adequate washing facilities.

4 Allowances

The following allowances are included in the Output Guide:

a. For contingencies and work other than that detailed in paragraph 2 an allowance of 21% has been made to the spraying, turn and prep spray times. Other work includes:
   (i)   general preparation;
   (ii)   adding chemical and water to tank in the ratio 1 : 1·5;
   (iii)   calibrating flow of chemicals through spray heads;
   (iv)   checking tank levels;
   (v)   adjusting arms to correct row width;
   (vi)   washing out pipes and spray heads (twice daily);
   (vii)   changing spray heads;
   (viii) to and from camp;
   (ix)   minor repairs of not more than 15 minutes accumulated duration during the day.

b. For personal needs and rest 18% of the total working time.

5 Method of using the Output Guide

a. Calculate the net area to be sprayed (total area less racks and rides).

b. Measure the mean distance between the rows.

c. Ascertain the mean length of the rows to be sprayed.

d. Select, from paragraph 6, the time to spray one hectare.

Example

12·6 hectares net area to be sprayed

2·2 m average distance between rows

260 m average row length

Time per hectare for 250 m row length = 58·59 standard minutes
(from para 6)

Time per hectare for 300 m row length = 57·65 standard minutes
(from para 6)

Time per hectare for 260 m row length

$$= 58·59 - \left[\frac{58·59 - 57·65}{5} \times 1 \right] \text{ (by interpolation)}$$

$$= 58·59 - 0·19$$

$$= 58·40 \text{ standard minutes}$$

N.B. To calculate the volume of chemical used, the formula is:

$$\text{Net area} \times \text{total rate of application per hectare} \times \frac{\text{swath width}}{\text{row spacing}}$$

In the example this would be:

$$12·6 \times 25 \times \frac{1·2}{2·2} = 172 \text{ litres of diluted chemical (this is made up of}$$

69 litres atrazine plus 103 litres of water).

Due allowance should be made for the quantity of chemical which may be lost during the operations of calibration, washing out pipes, turning at end of rows.

6 **Time in standard minutes to spray one hectare (including turn and prep spray) at a speed of 60 m per minute**

See table on page 94

7 **Modifications and variations to the Output Guide**

N.B. These modifications and variations are only to be applied when the prevailing conditions or job specification differ from those listed in paragraphs 1 and 2.

a. Should there be more than 10 obstructions per hectare it is advisable to spray using hand held equipment as a tractor mounted machine is not considered suitable.

b. Repairs of up to 15 minutes accumulated duration per day are included in the Output Guide, but any repair time in excess of 15 minutes should be paid for separately.

c. The weekly draining and washing out of the tank is not included in the Output Guide and should be carried out with the normal weekly tractor maintenance and be paid for separately.

Time in standard minutes to spray one hectare (including turn and prep spray) at a speed of 60 m per minute

Average distance between rows (m)	Average length of row in metres									
	50	75	100	150	200	250	300	350	400	450
2·2	75·51	68·46	64·70	61·41	59·53	58·59	57·65	57·18	56·71	56·71
2·3	72·21	65·16	61·87	58·58	56·70	55·76	55·29	54·82	54·35	53·88
2·4	69·11	62·53	59·24	55·95	54·54	53·60	52·66	52·19	52·19	51·72
2·5	66·19	60·08	56·79	53·97	52·09	51·15	50·68	50·21	49·74	49·74
2·6	63·89	57·78	54·96	51·67	50·26	49·32	48·85	48·38	47·91	47·91

Tractor mounted CDA—job description

The machine consists basically of a tank to hold the chemical, a pump, two solenoids and two 'Micron Herbi' spray heads.

The normal safety precautions should be observed when carrying out this operation. These may be checked by reference to:

a. FC Bulletin No 48.

b. The Entopath News Chemical Control Supplement.

N.B. The publications mentioned above are examples only and there may be other publications available to which reference may be made.

The tank should not be filled more than half way as considerable frothing of the atrazine occurs during spraying operations and may spill out of the filler cap.

1 **Preparing the machine for use**

    a. Add atrazine (Weedex 500L) and water to the tank in the ratio of 1 atrazine to 1·5 water. Always pour the liquids through the filter in the filler hole and always add atrazine before adding water. 2×5 litres of atrazine plus 3×5 litres of water should normally be sufficient for a morning's spraying (see paragraph 3e).

    b. Connect the outlet hose to the pump and the return hose to the tank (push and turn).

    c. Switch on the pump and both solenoids and wait until chemical emerges in a steady flow from each head.

    d. Calibrate by measuring the quantity of chemical that flows from each head in one minute. Use a measuring jar and a watch with a sweep second hand.

    The head may be tilted for this operation to concentrate the flow in one stream. The correct quantity is 180cc per minute from each head. Adjust the flow by turning the control valve (vertical one with alloy handle) clockwise to increase flow/anti-clockwise to reduce flow.

    If each head does not give the same flow, check that both heads are perfectly level and horizontal with each other (adjust by means of tractor's trailing arms), or that one nozzle is not partially blocked.

    e. Switch on spray heads and check that they are rotating satisfactorily. Leave running for one minute or so and observe size, shape and density of spray swath on ground. Correct size is 1·2m diameter, circular (oval indicates that wind may be too strong) and both circles should be of uniform density.

f. Switch sprayer off* and proceed to spraying area and position tractor midway between first two rows. Start from downwind side of compartment if the wind is across the rows.

g. Check the spray heads are over the centre of each row—adjusting arms accordingly.

h. Adjust height of spray heads to clear tops of trees.

2 Spraying

a. Switch the sprayer on* and check that spray is emerging.

b. Drive at a constant speed of 60 m per minute. Although individual tractors may vary, on the tractors studied, it was found that on the MF165 without Multi Power this speed was achieved by maintaining 1,200 rpm in third gear low. With Multi Power in high and also in third gear low 1,000–1,100 rpm was required. Obstacles such as stumps, hollows, etc, should be traversed with care and height of spray heads may need to be adjusted (raised) in order to avoid them grounding or hitting trees.

c. At the end of the row turn and check that spray heads are still working before commencing next row.

d. When driving to and from camp switch the sprayer off* and raise it on the tractor arms.

3 Washing out

When the machine is stopped for any reason for more than half an hour it is necessary to wash out the pipes and spray heads so that they will not become blocked with dried atrazine.

a. The tank outlet pipe and return pipe should be disconnected and the tank outlet pipe inserted into a container of clean water. If the stoppage is a temporary one during the day (lunch break, inclement weather) then no more than 5 litres of water need be used. At the end of the day 13–18 litres should be flushed through.

b. Switch sprayer on* and wait till clean water emerges from spray heads.

c. At the end of the day clean the grooves of the spray head discs with a cloth to remove all caked on atrazine.

d. Leave pipes disconnected until machine is ready to use again.

e. At the end of the day dilute any remaining chemical in the tank by adding some of the following day's water (10–15 litres).

f. Weekly, the tank should be drained and the inside washed to

*Five switches, one each for pump, solenoids and spray heads.

remove any dried on atrazine. The quickest way to drain the tank is to remove the return pipe connection and place this into a suitable container. Switch on the pump only and pump out the liquid.

After emptying and cleaning out the tank the pipes should be washed out as described above.

Note

When washing out pipes and tank due care must be taken to avoid run off into water courses.

Brashing Spruces

1. Conditions

The Output Guide applies to trees fully (100%) or partially (X%) brashed under the following conditions:

a. Slopes up to 35% (19°) with little or no undergrowth or rock outcrops.

b. Drains, where present, not closer than 6 m spacing.

c. Direct, turf or plough planting. Plough furrows up to 10 cm deep.

d. Average number of whorls to be brashed is 5·5 for NS and 5 for SS.

e. Average of 4·5 dead whorls above brashing height.

f. Few internodal branches.

N.B. **For conditions other than specified above see paragraph 7.**

2 Job specification

The Output Guide is for the following work:

a. Trees to be brashed to a height of 1·8 m (whorls occurring at exactly 1·8 m are not normally removed) measured from ground level (ie not from top of turf or from the bottom of a furrow). On slopes, brashing height is measured from side of the tree.

b. Branches to be cut close to the bole leaving no snags and not damaging the bark; this includes rideside and rough trees.

c. Forked, double and badly formed trees not to be brashed unless an unusually high proportion of the stand consists of this type of tree.

d. Trees with a DBH of under 7 cm not normally brashed.

e. Branches to be kept clear of racks, rides and main drains.

f. Partial brashing—pattern may be either lane or diffuse. With partial brashing, *all* trees must be brashed to the percentage chosen (but excluding trees in 2c and d above).

3. Tools and equipment (including safety equipment)

a. 0·5 m Skelton brashing saw with Skelton strapped head.

b. Handle 0·6 m long (preferably axe handle with fawn foot).

c. 15 cm three cornered slim taper file with handle.

d. 25 cm fine flat file with handle.

e. Eclipse No 77 setting pliers.

f. Filing vice.

g. Protective clothing viz:
   Safety helmet BS 2095 or 5240.
   Eye protection.
   Industrial gloves.
   Boots with safety toecap.

4 Allowances

Included in the times

a. For contingencies and work other than that performed on individual trees, eg sharpening, setting and cleaning the saw, walking between work sites etc, 10% of the time spent on actual brashing.

b. For personal needs and rest, 15% of the total working time.

5 Method of using the Output Guide

a. To select the time an assessment must be made of at least 100 trees. Normally it is desirable to set a rate for the job before the men begin work; with brashing this is not easy. Either the supervisor should brash 100 trees to make the assessment or the assessment should be made during the first day's work. It is important that this assessment should be made in average conditions for the stand. If necessary the assessment should be done in several parts of the stand to arrive at the coarseness index. In addition, the average number of whorls per tree should be checked.

The coarseness of each tree will be assessed as follows:

Fine branched: Under 13 mm diameter, brittle and more than one branch removed with one stroke of the saw.

Medium branched: 13–25 mm diameter, most branches removed with one stroke of the saw.

Coarse branched: Over 25 mm diameter, tough, most branches require more than 1 stroke of the saw. Sometimes many internodal branches.

b. *Calculation of the coarseness index*
   The number of trees in each category is expressed as a percentage of the total number counted. A multiplying factor (as shown on page 101) is then applied to each percentage.

	Norway spruce	Sitka spruce
Fine	3	3
Medium	4	4
Coarse	5	6

eg *Norway spruce*

Fine percentage by $\quad$ 3 eg $27\% \times 3 = 0\cdot81$
Medium percentage by 4 eg $56\% \times 4 = 2\cdot24$
Coarse percentage by $\quad$ 5 eg $17\% \times 5 = 0\cdot85$

Add these figures to give the coarseness index
$0\cdot81 + 2\cdot24 + 0\cdot85 = 3\cdot90$.

Sitka spruce

Fine percentage by $\quad$ 3 eg $27\% \times 3 = 0\cdot81$
Medium percentage by 4 eg $56\% \times 4 = 2\cdot24$
Coarse percentage by $\quad$ 6 eg $17\% \times 6 = 1\cdot02$

Add these figures to give the coarseness index
$0\cdot81 + 2\cdot24 + 1\cdot02 = 4\cdot07$.

c. *Selecting the time*

Look up the time per 100 trees for the species in paragraph 6 for the coarseness index and the percentage to be brashed.

eg NS Coarseness index $3\cdot90 = 117$ standard minutes per 100 trees 70% brashed.

SS Coarseness index $4\cdot07 = 113$ standard minutes per 100 trees 70% brashed.

6 Output Guides for brashing—Time in standard minutes per 100 trees

A *Norway Spruce*

Coarseness index	Percentage brashed and time per 100 trees					
	30%	*40%*	*50%*	*60%*	*70%*	*100%*
3·00	50	60	70	80	90	120
3·20	53	64	75	85	96	128
3·40	56	68	79	90	102	136
3·60	60	72	84	96	108	144
3·80	63	76	88	101	114	152
4·00	66	80	93	106	120	160
4·20	69	84	98	111	126	168
4·40	72	88	102	116	132	176
4·60	76	92	107	122	138	184
4·80	79	96	111	127	144	192
5·00	82	100	116	132	150	200

B *Sitka spruce*

Coarseness index	Percentage brashed and time per 100 trees					
	30%	*40%*	*50%*	*60%*	*70%*	*100%*
3·00	46	56	64	74	83	112
3·20	49	60	68	79	89	119
3·40	52	63	73	84	94	127
3·60	55	67	77	89	100	134
3·80	58	71	81	94	105	142
4·00	61	75	85	99	111	149
4·20	64	78	90	104	116	157
4·40	67	82	94	109	122	164
4·60	71	86	98	113	127	172
4·80	74	90	102	118	133	179
5·00	77	93	107	123	138	187
5·20	80	97	111	128	144	194
5·40	83	101	115	133	149	202
5·60	86	104	119	138	155	209
5·80	89	108	124	143	160	216
6·00	92	112	128	148	166	224

7 Modifications and variations to the Output Guide

N.B. These modifications and variations are only to be applied when the prevailing conditions or job specification differ from those listed in paragraphs 1 and 2.

a. *Better conditions*

Very easy going, almost level ground, screef or turf planted with no ground impediments and possibly had heavy weed growth in early years helping to suppress the lower branches.

Deduct in steps of 5% up to a maximum of 15%.

b. *Poorer conditions*

Hard going, deep (Cuthbertson) ploughing or slopes over 35%, probably broken ground with scrub and rock outcrops, intertwining branches.

Add in steps of 5% up to a maximum of 15%.

c. *Variation in number of whorls*

The times are for brashing an average 5·5 whorls in NS and 5 whorls in SS. If there is a difference

Add/subtract in steps of 5% per $\frac{1}{2}$ whorl.

d. *Variation in number of dead whorls above brashing height*

Where there are less than an average of 4·5 dead whorls above brashing height

Add 5%.

Where there are more than an average of 4·5 dead whorls above brashing height
Deduct 5%.

e. *Green branches*
Normally, green branches only occur on the edge of a stand and allowance for this is already made. When there is a significant proportion of green branches to be removed within the stand
Add 5% or 10%.

f. *Internodal branches*
Sometimes it is difficult to count whorls because of internodal branches. Depending on the occurrence
Add 5% or 10%.

Brashing Pines

1 Conditions

The Output Guide applies to trees fully (100%) or partially (X%) brashed under the following conditions:

a. Flat or gently sloping terrain up to 15% (8°) with occasional patches of undergrowth.

b. Few drains.

c. Direct or shallow plough planting.

d. Average number of whorls to be brashed is 5.

e. Average of 4·5 dead whorls above brashing height.

f. No internodal branches.

N.B. **For conditions other than specified above see paragraph 7.**

2 Job specification

The Output Guide is for the following work:

a. Trees to be brashed to a height of 1·8 m (whorls occurring at exactly 1·8 m are not normally removed) measured from ground level (ie not from top of plough ridge or from the bottom of a furrow). On slopes brashing height is measured from the side of of tree.

b. Branches to be cut close to the bole leaving no snags and not damaging the bark; this includes rideside and rough trees.

c. Forked, double and badly formed trees not to be brashed unless an unusually high proportion of the stand consists of this type of tree.

d. Trees with a DBH of under 5 cm not normally brashed.

e. Branches to be kept clear of racks, rides and main drains.

f. Partial brashing—pattern may be either lane or diffuse. With partial brashing, *all* trees must be brashed to the percentage chosen (but excluding trees in 2c and d above).

3 Tools and equipment (including safety equipment)

Saw brashing

a. 0·5m Skelton brashing saw with Skelton strapped head.

b. Handle 0·6m long (preferably axe handle with fawn foot).

c. 15cm three cornered slim taper file with handle.

d. 25cm fine flat file with handle.

e. Eclipse No 77 setting pliers.

f. Filing vice.

Billhook or axe brashing

g. Newton billhook (Spear & Jackson No 4501)
or

h. Yorkshire billhook with hooked end rounded off (Spear & Jackson No 4552)
or

i. 2lb trimming axe (Spear & Jackson No 3151).

j. Sharpening stone, carrying frog and edge tool gauge (see FC Bulletin No 48 Chapter 2).

k. Protective clothing viz:
Safety helmet BS 2095 or 5240.
Eye protection.
Industrial gloves.
Boots with safety toecap.

4 Allowances

Included in the times

a. For contingencies and work other than that performed on individual trees, eg sharpening, setting and cleaning the saw, billhook or axe, walking between work sites etc, 10% of the time spent on actual brashing.

b. For personal needs and rest 15% of the total working time.

5 Method of using the Output Guide

a. To select the time an assessment must be made of at least 100 trees.

Normally it is desirable to set a rate for the job before the men begin work; with brashing this is not easy. Either the supervisor should brash 100 trees to make the assessment or the assessment should be made during the first day's work. It is important that this assessment should be made in average conditions for the stand. If necessary the assessment should be done in several parts of the stand to arrive at the coarseness index. In addition the average number of whorls per tree should be checked.

The coarseness of each tree will be assessed as follows:

Fine branched. Under 13 mm diameter, brittle and more than one branch removed with one stroke of the saw, billhook or axe.

Medium branched. 13–25 mm diameter, most branches removed with one stroke of the saw, billhook or axe.

Coarse branched. Over 25 mm diameter, tough, most branches require more than one stroke of the saw, billhook or axe.

b. *Calculation of the coarseness index*

The number of trees in each category is expressed as a percentage of the total number counted. A multiplying factor (as shown below) is then applied to each percentage.

	Scots pine	Corsican pine	Lodgepole pine
Fine	4	3	3
Medium	5	4	4
Coarse	8	6	6

eg *Scots pine*

Fine   percentage by 4 eg $27\% \times 4 = 1 \cdot 08$
Medium percentage by 5 eg $56\% \times 5 = 2 \cdot 80$
Coarse   percentage by 8 eg $17\% \times 8 = 1 \cdot 36$
Add these figures to give the coarseness index
$1 \cdot 08 + 2 \cdot 80 + 1 \cdot 36 = 5 \cdot 24$

Corsican and Lodgepole pine

Fine   percentage by 3 eg $27\% \times 3 = 0 \cdot 81$
Medium percentage by 4 eg $56\% \times 4 = 2 \cdot 24$
Coarse   percentage by 6 eg $17\% \times 6 = 1 \cdot 02$
Add these figures to give the coarseness index
$0 \cdot 81 + 2 \cdot 24 + 1 \cdot 02 = 4 \cdot 07$

c. *Selecting the time*

Look up the time per 100 trees in paragraph 6 for the coarseness index, species and tool used and the percentage to be brashed.

eg Scots pine coarseness index $5 \cdot 24 = 48$ standard minutes per 100 trees 50% billhook or axe brashed and 64 standard minutes per 100 trees 50% saw brashed.

Corsican pine coarseness index $4 \cdot 07 = 45$ standard minutes per 100 trees 50% billhook or axe brashed and 59 standard minutes per 100 trees 50% saw brashed.

Lodgepole pine coarseness index $4 \cdot 07 = 38$ standard minutes per 100 trees 50% billhook or axe brashed and 52 standard minutes per 100 trees 50% saw brashed.

6 Output Guides for brashing—time in standard minutes per 100 trees

A. *Scots pine*

Billhook or axe						Coarse-ness index	Saw					
Percent brashed and time per 100 trees							Percent brashed and time per 100 trees					
30%	40%	50%	60%	70%	100%		30%	40%	50%	60%	70%	100%
26	32	37	42	47	63	4·00	35	41	49	56	63	84
27	34	39	44	50	66	4·20	37	43	51	59	66	88
29	35	41	46	52	69	4·40	39	46	54	62	69	92
30	37	43	48	54	72	4·60	40	48	56	64	73	97
31	38	44	50	57	76	4·80	42	50	59	67	76	101
33	40	46	53	59	79	5·00	44	52	61	70	79	105
34	42	48	55	61	82	5·20	46	54	64	73	82	109
35	43	50	57	64	85	5·40	47	56	66	76	85	113
36	45	52	59	66	88	5·60	49	58	68	78	88	118
38	46	54	61	68	91	5·80	51	60	71	81	92	122
39	48	56	63	71	94	6·00	53	62	73	84	95	126
40	50	57	65	73	98	6·20	54	64	76	87	98	130
42	51	59	67	75	101	6·40	56	66	78	90	101	134
43	53	61	69	78	104	6·60	58	68	81	92	104	139
44	54	63	71	80	107	6·80	60	70	83	95	107	143
46	56	65	74	82	110	7·00	61	73	86	98	110	147
47	58	67	76	85	113	7·20	63	75	88	101	114	151
48	59	68	78	87	117	7·40	65	77	90	104	117	155
49	61	70	80	89	120	7·60	67	79	93	106	120	160
51	62	72	82	92	123	7·80	68	81	95	109	123	164
52	64	74	84	94	126	8·00	70	83	98	112	126	168

B. Corsican pine

Billhook or axe						Coarse-ness index	Saw					
Percent brashed and time per 100 trees							Percent brashed and time per 100 trees					
30%	*40%*	*50%*	*60%*	*70%*	*100%*		*30%*	*40%*	*50%*	*60%*	*70%*	*100%*
24	28	33	38	43	57	3·00	32	38	44	50	57	76
26	29	35	40	46	61	3·20	34	40	46	53	60	81
27	31	37	43	49	65	3·40	36	43	49	56	64	86
28	33	39	46	52	68	3·60	38	45	52	60	68	91
30	35	41	48	54	72	3·80	40	48	55	63	72	96
32	37	44	51	57	76	4·00	43	50	58	66	76	101
33	39	46	53	60	80	4·20	45	53	61	70	79	106
35	41	48	56	63	84	4·40	47	55	64	73	83	111
36	42	50	58	66	87	4·60	49	58	67	76	87	116
38	44	52	61	69	91	4·80	51	60	70	80	91	121
40	46	55	63	72	95	5·00	53	63	73	83	95	126
41	48	57	66	74	99	5·20	55	65	76	86	98	131
43	50	59	68	77	103	5·40	58	68	79	90	102	136
44	52	61	71	80	106	5·60	60	70	82	93	106	141
46	54	63	73	83	110	5·80	62	73	85	96	110	146
48	56	66	76	86	114	6·00	64	76	88	100	114	152

C. Lodgepole pine

20	24	28	33	37	49	3·00	27	32	38	44	49	65
21	25	29	35	39	52	3·20	28	34	40	46	52	69
22	27	31	37	41	55	3·40	30	36	43	49	55	73
24	28	33	39	44	58	3·60	32	38	45	52	58	78
25	30	35	41	46	62	3·80	34	40	48	55	62	82
26	32	37	44	49	65	4·00	36	42	50	58	65	86
28	33	39	46	51	68	4·20	37	44	53	61	68	91
29	35	41	48	54	71	4·40	39	46	55	64	71	95
30	36	42	50	56	75	4·60	41	49	58	67	75	99
32	38	44	52	59	78	4·80	43	51	60	70	78	104
33	40	46	55	61	81	5·00	45	53	63	73	81	108
34	41	48	57	64	84	5·20	46	55	65	76	84	112
36	43	50	59	66	88	5·40	48	57	68	79	88	117
37	44	52	61	69	91	5·60	50	59	70	82	91	121
38	46	54	63	71	94	5·80	52	61	73	85	94	125
40	48	56	65	74	98	6·00	54	64	76	88	98	130

7 Modifications and variations to the Output Guide

N.B. These modifications and variations are only to be applied when the prevailing conditions or job specification differ from those listed in paragraphs 1 and 2.

a. *Poorer conditions*

Hard going, deep (Cuthbertson) ploughing or slopes over 15% (9°), frequent drains, broken ground, scrub and rock outcrops, intertwining branches.

Add in steps of 5% up to a maximum of 15%.

b. *Variation in number of whorls*

The times are for brashing an average 5 whorls per tree. If there is a difference

Add/subtract in steps of 5% per $\frac{1}{2}$ whorl.

c. *Variation in number of dead whorls above brashing height*

Where there are less than an average of 4·5 dead whorls above brashing height

Add 5%

Where there are more than an average of 4·5 dead whorls above brashing height

Deduct 5%

d. *Green branches*

Normally green branches only occur at the edge of a stand and allowance for this is already made. Where there is a significant proportion of green branches to be removed within the stand

Add 5% or 10%.

Brashing Japanese Larch

1 Conditions

The Output Guide applies to trees fully (100%) or partially (X%) brashed under the following conditions:

a. Slopes up to 35% (19°) with little or no undergrowth or rock outcrops.

b. Few drains.

c. Direct or shallow plough planting.

d. Whorls difficult to count but an average of about 4 to be brashed.

e. Branches dead for about 2 m above brashing height.

N.B. **For conditions other than specified above see paragraph 7.**

2 Job specification

The Output Guide is for the following work:

a. Trees to be brashed to a height of 1·8 m (whorls occurring at exactly 1·8 m are not normally removed) measured from ground level (ie not from top of turf or from the bottom of a furrow). On slopes brashing height is measured from the side of the tree.

b. Branches to be removed close to the bole leaving no snags and not damaging the bark; this includes rideside and rough trees.

c. Forked, double and badly formed trees not to be brashed unless an unusually high proportion of the stand consists of this type of tree.

d. Trees with a DBH of under 7 cm not normally brashed.

e. Branches to be kept clear of racks, rides and main drains.

f. Partial brashing—pattern may be either lane or diffuse. With partial brashing, *all* trees must be brashed to the percentage chosen (but excluding trees in 2c and d above).

3 Tools and equipment (including safety equipment)

Saw brashing

a. 0·5 m Skelton brashing saw with Skelton strapped head.

b. Handle 0·6m long (preferably axe handle with fawn foot).

c. 15cm three cornered slim taper file with handle.

d. 25cm fine flat file with handle.

e. Eclipse No 77 setting pliers.

f. Filing vice.

Stick brashing

g. Pick handle or locally cut stick.

h. Protective clothing viz:
   Safety helmet BS 2095 or 5240.
   Eye protection.
   Industrial gloves.
   Boots with safety toecap.

4 Allowances

Included in the times

a. For contingencies and work other than that performed on individual trees, eg sharpening, setting and cleaning the saw, walking between work sites etc 10% of the time spent on actual brashing.

b. For personal needs and rest 15% of the total working time.

5 Method of Using the Output Guide

a. To select the time an assessment must be made of at least 100 trees.

Normally it is desirable to set a rate for the job before the men begin work; with brashing this is not easy. Either the supervisor should brash 100 trees to make the assessment or the assessment should be made during the first day's work. It is most important that this assessment should be made in average conditions for the stand. If necessary the assessment should be done in several parts of the stand to arrive at the coarseness index. In addition the average number of whorls per tree should be checked.

The coarseness of each tree will be assessed as follows:

Fine branched. Under 13mm diameter, brittle and more than one branch removed with one stroke of the saw, pick handle or stick.

Medium branched. 13–25mm diameter, most branches removed with one stroke of the saw, pick handle or stick.

Coarse branched. Over 25mm diameter, tough, most branches require more than one stroke of the saw, pick handle or stick.

b. *Calculation of the coarseness index*

The number of trees in each category is expressed as a percentage of the total number counted. A multiplying factor (as shown below) is then applied to each percentage.

Fine   3
Medium 4
Coarse   6

eg Fine   percentage by 3 eg $27\% \times 3 = 0.81$
Medium   percentage by 4 eg $56\% \times 4 = 2.24$
Coarse   percentage by 6 eg $17\% \times 6 = 1.02$
Add these figures to give the coarseness index
$0.81 + 2.24 + 1.02 = 4.07$

c. *Selecting the time*

Look up the time per 100 trees in paragraph 6 for the coarseness index and the percentage to be brashed.

eg Coarseness index $4.07 = 43$ standard minutes per 100 trees 60% stick brashed and 65 standard minutes per 100 trees 60% saw brashed.

6 Output Guides for brashing—time in standard minutes per 100 trees

Japanese larch

Stick						Coarse-ness index	Saw					
Percent brashed and time per 100 trees							Percent brashed and time per 100 trees					
30%	40%	50%	60%	70%	100%		30%	40%	50%	60%	70%	100%
20	24	28	32	36	48	3·00	30	36	42	48	54	72
21	25	29	34	38	51	3·20	32	38	44	51	57	76
22	27	31	36	40	54	3·40	34	40	47	54	61	81
24	28	33	38	43	57	3·60	36	43	50	57	64	86
25	30	35	40	45	60	3·80	38	45	53	60	68	91
26	32	37	42	48	64	4·00	40	48	56	64	72	96
28	33	39	44	50	67	4·20	42	50	58	67	75	100
29	35	41	46	52	70	4·40	44	52	61	70	79	105
30	36	42	49	55	73	4·60	46	55	64	73	82	110
32	38	44	51	57	76	4·80	48	57	67	76	86	115
33	40	46	53	60	80	5·00	50	60	70	80	90	120
34	41	48	55	62	83	5·20	52	62	72	83	93	124
36	43	50	57	64	86	5·40	54	64	75	86	97	129
37	44	52	59	67	89	5·60	56	67	78	89	100	134
38	46	54	61	69	92	5·80	58	69	81	92	104	139
40	48	56	64	72	96	6·00	60	72	84	96	108	144

7 Modifications and variations to the Output Guide

N.B. These modifications and variations are only to be applied when the prevailing conditions or job specification differ from those listed in paragraphs 1 and 2.

a. *Better conditions*

Very easy going, almost level ground, few ground impediments and possibly had heavy weed growth in early years helping to suppress the lower branches

Deduct in steps of 5% up to a maximum of 15%.

b. *Poorer conditions*

Hard going, deep (Cuthbertson) ploughing or slopes over 35% (19°), probably broken ground with scrub and rock outcrops, intertwining branches

Add in steps of 5% up to a maximum of 15%.

c. *Variation in number of whorls*

Where the average number of whorls brashed is more or less than 4

Add/subtract in steps of 5% per $\frac{1}{2}$ whorl

d. *Variation in height of dead branches above brashing height*

Where dead branches extend for less than an average of 2 m above brashing height

Add 5%

Where dead branches extend for more than an average of 2 m above brashing height

Deduct 5%

e. *Green branches*

Normally, green branches only occur at the edge of a stand and allowance for this is already made. Where there is a significant proportion of green branches to be removed within the stand

Add 5% or 10%

Brashing Douglas Fir and Grand Fir

Section	XVI
No	G4

1 Conditions

The Output Guide applies to trees fully (100%) or partially (X%) brashed under the following conditions:

a. Slopes up to 35% (19°) with occasional patches of undergrowth and scrub and some rock outcrops.

b. Few drains.

c. Direct, turf or shallow plough planting.

d. Average number of whorls to be brashed is 4.

e. Average of 5·5 dead whorls above brashing height in Douglas fir and 2 in Grand fir.

f. Few internodal branches.

N.B. **For conditions other than specified above see paragraph 7.**

2 Job specification

The Output Guide is for the following work:

a. Trees to be brashed to a height of 1·8 m (whorls occurring at exactly 1·8 m are not normally removed) measured from ground level (ie not from top of turf or from the bottom of a furrow). On slopes the brashing height is measured from side of tree.

b. Branches to be cut close to the bole leaving no snags and not damaging the bark (extra care to be taken with Grand fir); this includes rideside and rough trees.

c. Forked, double and badly formed trees not to be brashed unless an unusually high proportion of the stand consists of this type of tree.

d. Trees with a DBH of under 7 cm not normally brashed.

e. Branches to be kept clear of racks, rides and main drains.

f. Partial brashing—pattern may be either lane or diffuse. With partial brashing, *all* trees must be brashed to the percentage chosen (but excluding trees in 2c and d above).

115

3 Tools and equipment (including safety equipment)

    a. 0·5 m Skelton brashing saw with Skelton strapped head.

    b. Handle 0·6 m long (preferably axe handle with fawn foot).

    c. 15 cm three cornered slim taper file with handle.

    d. 25 cm fine flat file with handle.

    e. Eclipse No 77 setting pliers.

    f. Filing vice.

    g. Protective clothing viz:
       Safety helmet BS 2095 or 5240.
       Eye protection.
       Industrial gloves.
       Boots with safety toecap.

4 Allowances

Included in the times

    a. For contingencies and work other than that performed on individual trees, eg sharpening, setting and cleaning the saw, walking between work sites etc 10% of the time spent on actual brashing.

    b. For personal needs and rest 15% of the total working time.

5 Method of using the Output Guide

    a. To select the time an assessment must be made of at least 100 trees.

Normally it is desirable to set a rate for the job before the men begin work; with brashing this is not easy. Either the supervisor should brash 100 trees to make the assessment or the assessment should be made during the first day's work. It is most important that this assessment should be made in average conditions for the stand. If necessary the assessment should be done in several parts of the stand to arrive at the coarseness index. In addition the average number of whorls per tree should be checked.

The coarseness of each tree will be assessed as follows:

Fine branched. Under 13 mm diameter, brittle and more than one branch removed with one stroke of the saw.

Medium branched. 13–25 mm diameter, most branches removed with one stroke of the saw.

Coarse branched. Over 25 mm diameter, tough, most branches require more than one stroke of the saw. Sometimes many internodal branches.

    b. *Calculation of the coarseness index*
    The number of trees in each category is expressed as a percentage of the total number counted. A multiplying factor is then applied to each percentage.

	Douglas fir	Grand fir
Fine	4	3
Medium	5	4
Coarse	7	6

eg *Douglas fir*

Fine percentage by   4 eg $27\% \times 4 = 1{\cdot}08$
Medium percentage by 5 eg $56\% \times 5 = 2{\cdot}80$
Coarse percentage by   7 eg $17\% \times 7 = 1{\cdot}19$
Add these figures to give the coarseness index
$1{\cdot}08 + 2{\cdot}80 + 1{\cdot}19 = 5{\cdot}07$.

Grand fir

Fine percentage by   3 eg $27\% \times 3 = 0{\cdot}81$
Medium percentage by 4 eg $56\% \times 4 = 2{\cdot}24$
Coarse percentage by   6 eg $17\% \times 6 = 1{\cdot}02$
Add these figures to give the coarseness index
$0{\cdot}81 + 2{\cdot}24 + 1{\cdot}02 = 4{\cdot}07$.

c. *Selecting the time*

Look up the time per 100 trees in paragraph 6 for the coarseness index, species and the percentage to be brashed.

Eg Douglas fir coarseness index $5{\cdot}07 = 89$ standard minutes per 100 trees 60% brashed.

Grand fir coarseness index $4{\cdot}07 = 103$ standard minutes per 100 trees 60% brashed.

6 Output Guides for brashing—time in standard minutes per 100 trees

A   *Douglas fir*

Coarseness index	Percentage brashed and time per 100 trees					
	30%	*40%*	*50%*	*60%*	*70%*	*100%*
4·00	44	53	62	70	79	106
4·20	46	56	65	73	83	111
4·40	48	58	68	77	87	117
4·60	51	61	71	81	91	122
4·80	53	63	74	84	95	127
5·00	55	66	77	88	99	133
5·20	57	69	80	91	103	138
5·40	59	71	83	95	107	143
5·60	62	74	87	98	111	149
5·80	64	76	90	102	115	154
6·00	66	79	93	105	119	159
6·20	68	82	96	109	123	165
6·40	70	84	99	112	127	170
6·60	73	87	102	116	131	175
6·80	75	89	105	119	135	181
7·00	77	92	108	123	139	186

B *Grand fir*

Coarseness index	Percentage brashed and time per 100 trees					
	30%	*40%*	*50%*	*60%*	*70%*	*100%*
3·00	47	57	67	76	86	115
3·20	50	60	71	81	91	123
3·40	53	64	75	86	97	130
3·60	56	68	80	91	103	138
3·80	59	72	84	96	108	146
4·00	62	76	89	101	114	153
4·20	65	79	93	106	120	161
4·40	68	83	98	111	126	169
4·60	72	87	102	116	131	176
4·80	75	91	107	121	137	184
5·00	78	95	111	126	143	192
5·20	81	98	116	131	149	199
5·40	84	102	120	136	154	207
5·60	87	106	125	141	160	215
5·80	90	110	129	146	166	222
6·00	94	114	134	152	172	230

7 Modifications and variations to the Output Guide

N.B. These modifications and variations are only to be applied when the prevailing conditions or job specification differ from those listed in paragraphs 1 and 2.

a. *Better conditions*

Very easy going, almost level ground, screef or turf planted with no ground impediments and possibly had heavy weed growth in early years helping to suppress the lower branches

Deduct in steps of 5% up to a maximum of 15%.

b. *Poorer conditions*

Hard going, deep (Cuthbertson) ploughing or slopes over 35% (19°), probably broken ground with scrub and rock outcrops, intertwining branches

Add in steps of 5% up to a maximum of 15%.

c. *Variation in number of whorls*

The times are for brashing an average 4 whorls. If there is a difference

Add/subtract in steps of 5% per $\frac{1}{2}$ whorl.

d. *Variation in number of dead whorls above brashing height*

Douglas fir

Where there are less than an average of 5·5 dead whorls above brashing height

Grand fir

Where there are less than an average of 2 dead whorls above brashing height

   Add 5%.

Douglas fir

Where there are more than an average of 5·5 dead whorls above brashing height

Grand fir

Where there are more than an average of 2 dead whorls above brashing height

   Deduct 5%.

e. *Green branches*

Normally, green branches only occur on the edge of a stand and allowance for this is already made. Where there is a significant proportion of green branches to be removed within the stand

   Add 5% or 10%.

Marking and Tariffing Pines and Spruces Selective Thinning, Line Thinning and Clear Felling

1 Conditions

The times apply to crops marked and tariffed under the following conditions:

a. Complete brashing (see para 7a).

b. Trees free of ivy and/or similar climbers (see para 7b).

c. Tariffing area reasonably compact (see para 7c).

d. Slopes of up to 25% (14°) (see para 7d).

2 Job specification

The general procedure and necessary conventions for timber measurement are given in FC Booklet No 36. It is essential that the various measuring conventions are strictly adhered to.

It is important that the techniques shown below are followed if efficient working is to be achieved and the Output Guide to apply.

(i) *Marking*

Where a considerable percentage of the crop is unbrashed a long handled slasher should be used for marking. This can usually be pushed between the branches and the tree marked without any need to 'brash' first.

In areas that have tall vegetation (or will have before the tariffing or felling is completed) the blaze marks should be made as high up the tree as is practical.

(ii) *Girthing*

The girthing tape should be held in the hand between girth samples and not put in a pocket or draped around the neck.

(iii) *Identifying volume sample trees*

When the volume sample trees are cut as a separate operation it is important that a clear, quick means of identifying the tree is used. The usual method of blazing the trees can be a very lengthy process and often not easily seen later. Coloured plastic tape tied around the trunk is far quicker and more effective.

In line thinning, particularly if unbrashed, much time can be saved by identifying on row end trees those rows containing volume sample trees and the numbers involved.

Where volume sample trees are being cut as a separate operation the booker should always make a sketch of the area, marking in the rough position of the samples in relation to some identifiable point.

(iv) *Work on volume sample trees*
Only sufficient snedding to allow accurate measurement of the tree should be carried out.

While the feller is snedding the sample the booker should treat the stump and run out the tape. When the snedding is completed the booker scribes the 7 cm diameter length and mid length point. The feller then walks back down the tree with his saw and measures and calls out the mid diameter point as he passes.

(v) *Team composition*
A marking and tariffing team consists of a booker together with a number of markers and possibly additional men felling volume sample trees.

(vi) The most efficient team size in any given situation is:

a. where the booker is working with the highest number of markers that he can accurately control

b. where the team does the minimum amount of walking over the area commensurate with the minimum of lost time by the team as a whole.

(vii) The number of markers per booker will depend on the marking conditions; ie whether a clear felling, selective or line thinning, density of marking, ease of access etc. The minimum team size which eliminates the possibility of bias in selection of girth and volume sample trees is one booker and two markers. Generally, efficient, accurate working cannot be carried out with a larger team than one booker and five markers. As a general recommendation team size will be:

2 markers and 1 booker in first thinning.

3 or 4 markers and 1 booker in second or subsequent thinnings.

In clear felling team size will depend on the density of the stand and the number of blaze marks made per tree.

(viii) *Working method*
Walking will be minimised when the marking and girthing, felling and measuring are all carried out in one operation as the team proceeds through the stand.

In early thinning (where the volume samples are small, can easily be cut down with a bowsaw, gang size is small and access and visibility poor due to lack of brashing) this method of working is often the most suitable. At least one marker should use a fairly heavy marking tool—eg light axe or Yorkshire billhook—which can then be used for snedding out the volume sample trees.

(ix) In later thinning and clear felling, when a chainsaw will save considerable time on the volume sample trees, it is usually more efficient to fell the samples as a separate operation.

Ideally in this situation the felling and measuring of samples should be carried out by a team of two, at least one of whom was in the original marking team and will know the marked area and rough position of the samples.

The times are for the following work:

a. *Selective thinning*
  (i) Select the trees to be removed and blaze at two points on the tree, either above or below the 1·3 m height, as appropriate to the conditions. (For more or less than two blaze marks per tree see paragraph 7e.)
  (ii) Measure the girth at 1·3 m height on girth sample trees. (Where the use of a height measuring stick and/or scribe mark are required see paragraph 7f.)
  (iii) On volume sample trees scribe girthing point, number and identify, by means of plastic tape or similar, for later felling and measuring.

N.B. No identification required where trees are felled at the time of marking and tariffing.

  (iv) Search for and fell volume sample trees. Sned sufficiently for accurate measurement of length to 7 cm top diameter, and mid diameter, scribing the two points.

N.B. No search time where trees are felled at the time of marking and tariffing.

  (v) Book all necessary data for calculation of the tariff volume on Form FC U15 (see Booklet 36).

N.B. The time for calculation of tariff volume is *not* included in the Output Guide.

b. *Line thinning*
  (i) Select appropriate rows to be removed and blaze each tree at one point below the 1·3 m height (for variations in blazing see paragraph 7g).
  (ii) Remainder of job specification as for selective thinning.

c. *Clear felling*
  (i) Blaze the trees at two points above or below the 1·3 m height as appropriate to the conditions. (For more or less than two blaze marks see paragraph 7e.)
  (ii) Remainder of job specification as for selective thinning.

3 Tools and equipment (including safety equipment)

a. *Samples cut at time of marking and tariffing*
Each marker:
Suitable marking tool.
Metric diameter tape.
Pocket scribe.

(N.B. In certain situations a 1·3 m long measuring rod may be required.)

Booker:
FC U15 Forms (see page 17 of FC Booklet No 36).
Board and ball point pen.
Metric diameter tape.
Metric 20 m tape.
Bowsaw.
Stump treatment liquid and container.

b. *Samples cut as a separate operation*
Each marker:
As in 3a above plus plastic tape for sample tree identification.
Booker:
As in 3a above less bowsaw and stump treatment liquid.

c. *Cutting and measuring volume sample trees as a separate operation*
Lightweight anti-vibration chainsaw complete with spares and fuel.
Breaking bar/cant hook.
Plastic wedge.
Stump treatment liquid and container.
FC U15 Forms (see page 17 of FC Booklet No 36).
Board and ball point pen.
Two metric diameter tapes.
One metric 20 m tape.

d. *Protective clothing*
Felling only:
Safety helmet BS 2826, 2095 or 5240.
Eye protection.
Ear defenders.
Gloves, incorporating ballistic nylon lining on back of left hand.
Nylon leg guard.
Safety boots, incorporating inner ballistic nylon guard.

Marking and tariffing only:
Safety helmet BS 2826, 2095 or 5240.
Eye protection.
Boots with safety toecap.

4 Allowances

The following allowances are included in the times:

a. For contingencies and other work.

    (i)   On marking and tariffing for work other than that performed on individual trees, eg discussion, move rows, move blocks, to and from camp etc 26% of the time spent on marking and tariffing.

    (ii) On walk and search, fell and measure volume sample trees, for work other than that performed on individual trees eg tools upkeep, discussion, move block etc 13% of the time spent on volume sample trees.

b. For personal needs and rest 12–22% of the total working time.

5 Method of using the Output Guide

According to the crop and marking situation, team composition and working method, select the various time elements concerned from paragraph 6. By multiplication and addition arrive at an overall time for the team.

N.B. Access score (graded 1 easy, 2 moderate and 3 difficult) is a subjective assessment of the difficulty of walking through the area, taking into account all ground and crop conditions except slope, which is accounted for separately in paragraph 7d.

Example

Selective thinning in Scots pine:

3 men marking and tariffing and 1 man booking.

2 men felling the samples later as a separate operation.

Mean sample tree volume $0.08\,m^3$.

500 stems marked per hectare.

Access score 1.

Girth sample tree fraction 1 in 10.

Volume sample tree fraction 1 in 100.

a. Marking and tariffing:

Element	Time SMs	Total time SMs/Ha
Walk and consider (from paragraph 6A)	104·00	104·00
Mark per 10 trees (from paragraph 6D)	0·84 × 50	42·00
Girth per 10 trees (1 in 10 girthing fraction) (from paragraph 6E)	1·93 × 5	9·65

Scribe, number and
identify volume sample
trees (1 in 100 sample
fraction)
(from paragraph 6F) 1.68×5 $\underline{8.40}$

$\overline{164.05}$

Multiply this total time by $\dfrac{\text{Number in team}}{\text{Number of markers}}$

ie $164.05 \times \dfrac{4}{3} = 218.7$ standard minutes.

b. Finding, cutting and measuring volume sample trees:

Element	Time SMs × trees/Ha × Men	Total time SMs/Ha
Walk and search (access score 1, 5 samples/Ha) (from paragraph 6G)	$1.74 \times 5 \times 2$	17.40
Fell and measure volume sample tree (pine $0.08 \, \text{m}^3$) (from paragraph 6H)	$4.88 \times 5 \times 2$	$\underline{48.80}$
	Total	$\overline{66.20}$

Total time in standard minutes per hectare for marking, tariffing and cutting volume sample trees $= 218.7 + 66.20 = 284.9$ standard minutes.

6A Selective thinning—'walk and consider' time in standard minutes per hectare

Stems per hectare	Pines			Spruces		
	Access score 1	Access score 2	Access score 3	Access score 1	Access score 2	Access score 3
50	27	31	35	18	22	26
100	49	56	64	35	43	50
150	65	76	87	50	62	73
200	77	92	107	64	79	94
250	86	104	123	76	95	113
300	91	114	136	87	110	132
350	95	121	147	97	123	149
400	98	128	158	106	136	166
450	101	134	168	114	148	181
500	104	141	178	121	158	196
550	109	149	190	128	169	210
600	115	160	204	134	178	223
650	124	173	221	139	188	236
700	134	186	238	145	197	249

6B Line thinning—'walking time' in standard minutes per hectare

Access score	1			2			3		
Row spacing	1·5m	1·8m	2·1m	1·5m	1·8m	2·1m	1·5m	1·8m	2·1m
1 row in 4 marked	41	34	29	53	44	38	65	54	46
1 row in 3 marked	54	45	39	70	58	50	86	72	62

6C Clear felling—'walking time' in standard minutes per hectare

Stems per hectare	Access score 1	Access score 2	Access score 3
100	24	32	39
200	34	45	55
300	42	55	67
400	49	63	77
500	55	71	87
600	60	77	95
700	64	84	102

6D *Selective thinning, line thinning and clear felling—'marking' time in standard minutes per 10 trees.*
2 blaze marks per tree, 0·84 standard minutes per **10** trees.

6E *Selective thinning, line thinning and clear felling—'girthing' time in standard minutes per 10 trees.*
1·93 standard minutes per **10** trees.

6F *Selective thinning, line thinning and clear felling—'scribe, number and identify' volume sample trees time in standard minutes per tree.*
Pines   1·68 standard minutes per tree.
Spruces 1·25 standard minutes per tree.

6G Selective thinning, line thinning and clear felling—time to 'walk and search' for volume sample trees in standard minutes per tree per man

Volume sample trees per hectare	Access score 1	Access score 2	Access score 3
1	3·89	5·19	6·48
2	2·75	3·67	4·59
3	2·24	3·00	3·75
4	1·95	2·59	3·24
5	1·74	2·32	2·91
6	1·58	2·11	2·65
7	1·47	1·96	2·45
8	1·37	1·84	2·30
9	1·30	1·72	2·17
10	1·23	1·63	2·05
12	1·13	1·50	1·88
14	1·04	1·39	1·74
16	0·97	1·30	1·62
18	0·92	1·22	1·53
20	0·87	1·17	1·45

6H Selective thinning, line thinning and clear felling—time to 'fell and measure' volume sample trees in standard minutes per tree per man

Tree volume m³	Pines	Spruces
0·02	4·07	4·35
0·04	4·35	4·88
0·06	4·62	5·38
0·08	4·88	5·85
0·10	5·13	6·29
0·12	5·36	6·75
0·14	5·58	7·09
0·16	5·79	7·46
0·18	5·99	7·80
0·20	6·18	8·12
0·22	6·35	8·42
0·24	6·51	8·70
0·26	6·66	8·96
0·28	6·80	9·21
0·30	6·93	9·45
0·32	7·04	9·67
0·34	7·14	9·89
0·36	7·23	10·10
0·38	7·31	10·30
0·40	7·38	10·49

7 Modifications and variations to the times

N.B. These modifications and variations are only to be applied when the prevailing conditions or job specification differ from those listed in paragraphs 1 and 2.

a. *Reduced intensity brashing*

(i) *Marking*

Where possible a suitable tool (eg long handled slasher) should be used to enable marking to take place without prior brashing of the trees. Where this is not possible the times shown in paragraph 7a(ii) below replace that in paragraph 6D.

(ii) *Girthing*

The times shown below replace that shown in paragraph 6E when the brashing percentage is less than 100%.

Time in standard minutes per 10 trees		
Brashing percentage	*Marking*	*Girthing*
90	0·98	2·03
80	1·11	2·13
70	1·25	2·23
60	1·39	2·33
50	1·53	2·43
40	1·67	2·53
30	1·81	2·63

(iii) *Scribing, numbering and identifying volume sample trees*

The times shown below replace that shown in paragraph 6f when the brashing percentage is less than 100%.

Brashing percentage	*Time in standard minutes per tree*	
	Pine	*Spruce*
90	1·72	1·29
80	1·76	1·33
70	1·80	1·37
60	1·84	1·41
50	1·88	1·45
40	1·92	1·49
30	1·96	1·53

N.B. *Line thinning*

In line thinning different methods of brashing will have varying effects on marking etc eg a $33\frac{1}{3}$% brashing, where the trees are completely brashed in the rows to be marked (1 line in 3) is equivalent to a 100% brashing with regard to marking, girthing etc and no modification is required to the times.

b. *Trees with ivy or similar climbers*
    (i) Where it is necessary to cut climbers from the tree before it can be marked,
    Add 0·05 or 0·10 standard minutes per tree involved according to the density of the climber.
    (ii) Where it is necessary to remove a climber on the girth sample tree to facilitate girthing,
    Add from 0·10–0·20 standard minutes in steps of 0·05 standard minutes per tree involved according to the density of the climber.
    N.B. **This replaces the addition given in paragraph 7b(i) above for these trees.**

c. *Tariffing area composed of small scattered blocks*
    Add from 5–10% to the times.

d. *Steep slopes*
    Where slopes are steeper than 25% (14°) increase the times as follows:

Slope		Increase in time
26–35%	(15°–19°)	5%
36–45%	(20°–24°)	10%
46% +	(25° +)	15%

e. *More or less than two blaze marks per tree*
    The times shown below replace that shown in paragraph 6D.
    1 blaze mark per tree —0·42 standard minutes per **10** trees.
    3 blaze marks per tree—1·26 standard minutes per **10** trees.
    4 blaze marks per tree—1·68 standard minutes per **10** trees.

f. *Use of a height measuring stick and/or scribe mark when girthing*
    Where the job specification stipulates:
    (i) the use of a height measuring stick for determining the girthing point,
    and/or
    (ii) that a scribe mark is made at the girthing point,
    Add 0·06 standard minutes to the girthing time for each operation.

g. *Only row end trees marked in a line thinning*
    Where the job specification requires that only 3–4 trees are marked at the ends of each line being removed, give no 'mark' time (paragraph 6D) but add 0·40 standard minutes in brashed stands and 0·88 standard minutes in unbrashed stands per marked row.

h. *Sample plots*

Where small sample plots (usually 0·01 or 0·05 hectares) are required to be put in by the tariffing team, the following additions should be made.

(i) Plots for assessing stems per hectare being marked to obtain correct girthing fraction

10 standard minutes per plot per man.

(ii) Plots for controlling volume being marked

6 standard minutes per plot per man.

Standard Times for Thinning Corsican and Scots Pine Using a Chainsaw

1 Conditions

The standard times apply to stands thinned under the following conditions:

a. The volume of the average tree is determined by the tariff system (see FC Booklet No 36).

b. The brashing percentage is at least 90%.

c. Forest floor conditions are average, ie generally flat or with a slope of not more than 10% (6°) and with some decayed brash and undergrowth. For worse conditions see para 7.

d. Racks are 40 m apart.

e. Supplies of chainsaw fuel and oil stump treatment solution are readily available at convenient places.

2 Job specification

The standard times are for the following work:

a. Trees to be felled in the direction as instructed by the supervisor.

b. Trees to be cut low, stumps not more than 0·025 m above ground level.

c. Branches and prominent knots to be cut off flush with the wood so that no snags remain. Snedding to be by chainsaw.

d. Felling and other cuts to be made squarely across the tree. Butts must be trimmed square if necessary.

e. Lop and top to be cleared from racks and rides unless otherwise directed by the supervisor.

f. Stumps to be treated immediately after felling. (This work attracts extra standard time—see para 7b.)

g. See Appendix for method of working.

3 **Tools and equipment** (including safety equipment)
   a. Lightweight self oiling chainsaw of a type suitable for snedding, with spares and maintenance tools.

   b. A 0·13 m plastic felling wedge.

   c. A pulp hook.

   d. A cant hook.

   e. Protective clothing viz:
   Safety helmet BS 2826, 2095 or 5240.
   Eye protection.
   Ear defenders.
   Gloves, incorporating ballistic nylon lining on back of left hand.
   Nylon leg guard.
   Safety boots, incorporating inner ballistic nylon guard.

4 **Allowances**

   Included in the standard times
   a. For contingencies and work other than that performed on individual trees, eg refuelling and day-to-day maintenance of the saw, clearing brash from rides etc. 20% of the time spent on actual thinning.

   b. For personal needs and rest: 20% of the total working time.

   To be included in the price per standard minute
   c. The appropriate incentive allowance.

   Chainsaw usage when thinning and chainsaw snedding is 60%.

5 **Method of selecting the standard time**
   a. The trees shall be tariffed according to the method specified in FC Booklet No 36. At the same time the number of whorls of branches to be cut off between butt and the 0·07 m top diameter over bark point on the felled sample trees shall be recorded. Where the felled sample tree has been pruned, only the whorls remaining to be cut off shall be counted.

   b. To select the standard time from table 6A (Scots pine) or 6B (Corsican pine) select the appropriate average volume for the stand (rounding up or down to the nearest) and read off the standard time from col (ii). This time is for the average number of whorls shown in col (iii). If the number of whorls is greater or less than this figure, add/subtract the standard minutes per whorl as shown at the foot of the table. If the average number of whorls is outside the range shown in col (iv), consult the local Work Study Officer.

Example
Thinning Scots pine
Mean tree volume 0·100
Mean number of whorls per tree = 15
Mean stump diameter 0·42 m
Conditions and job specification as shown in paragraphs 1 and 2

Standard time per tree (from 6A)	= 4·04 SMs
Addition for 2 extra whorls	= 0·26 SMs
Allowance for stump treatment (from 7b)	= 0·42 SMs
Total time per tree	= 4·72 SMs

6A Standard times for thinning in Scots pine (chainsaw snedding)

Average tree (cubic metres) (i)	Standard mins per tree (ii)	Average no of whorls (iii)	Range of whorls (iv)
0·036	2·95	10	7–14
0·045	3·23	11	7–15
0·065	3·55	12	8–16
0·080	3·71	12	8–16
0·100	4·04	13	9–17
0·115	4·29	14	10–18
0·145	4·62	14	10–18
0·170	4·98	15	11–19
0·190	5·27	15	11–19
0·225	5·75	16	12–20
0·250	6·07	17	13–21
0·280	6·52	18	14–21
0·315	7·08	19	15–22
0·350	7·56	20	16–23
0·395	8·24	22	18–25
0·440	8·86	23	19–26
0·470	9·33	24	20–28

Add/Subtract 0·13 SMs per whorl if different from average value given in col (iii) but do not go beyond range given in col (iv).

6B Standard times for thinning in Corsican pine (chainsaw snedding)

Average tree (cubic metres) (i)	Standard mins per tree (ii)	Average no of whorls (iii)	Range of whorls (iv)
0·110	5·77	16	11–21
0·130	6·00	16	11–22
0·160	6·26	17	12–22
0·180	6·45	17	12–22
0·220	6·79	18	13–23
0·250	7·12	18	13–23
0·305	7·47	18	13–23
0·350	7·85	19	14–24
0·395	8·26	19	14–24
0·460	8·96	20	15–25
0·540	9·68	21	15–25
0·575	9·91	21	16–26
0·620	10·34	22	17–27
0·665	10·82	22	17–27
0·720	11·24	22	17–27

Add/Subtract 0·25 SMs per whorl if different from the average value given in col (iii) but do not go beyond the range given in col (iv).

7 Modifications and variations to the standard times

N.B. These modifications and variations are only to be applied when the prevailing conditions or job specification differ from those listed in paragraphs 1 and 2.

a. *Presence of bramble*

Where conditions are more difficult because of the presence of bramble, an addition may be necessary. The average height of the bramble throughout the area for which the rate is being calculated should be assessed and the additions shown below added to *all* trees. (Parts of the area may have no bramble and must be allowed for.)

Mean bramble height (m)	Add per tree (SMs)
0·3	0·41
0·6	0·49
0·9	0·58
1·2	0·66
1·5	0·74
1·8	0·82
2·1	0·90
2·4	0·99
2·7	1·07

b. *Stump treatment*

The standard times include:

Preparation (making up the suspension in the case of *Peniophora*); filling the container, applying the liquid to cover completely all the exposed wood of the stumps that result from felling the tree immediately after felling, and an allowance for working with the dye. The following standard times may be added to tables 6A or 6B or paid separately as a price per stump:

Approximate average stump diameter	*Standard minutes per stump*
Up to 0·15m	0·10
Over 0·15m and up to 0·28m	0·20
Over 0·28m and up to 0·40m	0·32
Over 0·40m and up to 0·53m	0·42
Over 0·53m and up to 0·66m	0·52
Over 0·66m and up to 0·79m	0·62
Over 0·79m and up to 0·91m	0·72

c. *Tops*

When trees are snedded beyond the 0·07m top diameter-over-bark-point, additions to the standard times are necessary as shown in the table below. The extra time for the *average* cut off diameter should be added to *all* trees.

Top diameter OB	*Standard time*
0·06m	0·55
0·05m	0·67
0·04m	0·82
0·03m	0·96
0·02m	1·10

Method of working

The recommended method of felling and snedding trees by chain-saw appropriate to this table is as follows:

(i)   Walk to tree carrying saw, cant hook, stump treatment bottle and saw maintenance tools, logger's tape and pulp hook in belt.

(ii)   Prepare to fell by removing brash, deep litter and vegetation from around stump; start saw.

(iii)   Fell tree in the required direction.

†(iv)   Take down, using cant hook if necessary (throw cant hook).

*(v)   Trim butt.

*(vi)   Treat stump.

(vii)   Place logger's tape hook on tree butt.

(viii)   Sned to approximate point for crosscutting sawlog.

(ix)   Check measurement with tape, release tape, crosscut log.

†(x)   Turn log (throw cant hook).

(xi)   If necessary, re-hook tape to measure further length.

(xii)   Sned to top, off top.

(xiii)   Turn pole length from tip.

(xiv)   Sned back through various lengths to butt end.

*(xv)   Trim butt and treat stump (if not done at (v) and (vi)).

(xvi)   Collect tools and walk to next tree.

Notes:

*If feller has to go all the way back to the stump to sned it, it is better to treat stump last. It may also be better to trim butt last when the tree has been turned.

†In big trees, where it is necessary to cut large logs which cannot be turned by hand, it is better to throw the cant hook progressively forward to each crosscutting position and turn each piece after it has been crosscut. It is a matter of experience to know when the cant hook is likely to be needed.

1 Conditions

The standard times apply to stands thinned under the following conditions:

a. The average volume per tree determined by the 'tariff' system (see FC Booklet No 36).

b. The brashing percentage at least 90, but see paragraph 7a.

c. The floor conditions are average for the species and locality, but see paragraph 7c.

2 Job specification

The standard times are for the following:

a. Stumps to be cut as low as possible.

b. Branches and pruning knots to be cut off flush with the stem of the tree. Snedding by chainsaw.

c. Tops to be cut off at a diameter of 7 cm unless otherwise instructed by the supervisor.

d. Lop and top to be cleared from racks, roads and main drains.

e. All cuts including the felling cut, to be made as squarely across the tree as possible.

f. The supervisor will stipulate the direction of extraction and all trees should be felled in a direction to assist the subsequent extraction.

g. All buttressing to be removed from butt.

3. Tools and equipment (including safety equipment)

a. Chainsaw (lightweight) suitable for chainsaw snedding spares and tools.

b. Plastic bottle with brush for stump treatment.

c. Canthook, felling tongs or breaking bar.

d. Small alloy wedge.

e. Protective clothing viz:
   Safety helmet BS 2826, 2095 or 5240.
   Eye protection.
   Ear defenders.
   Gloves, incorporating ballistic nylon lining on back of left hand.
   Nylon leg guard.
   Safety boots, incorporating inner ballistic nylon guard.

4 Allowances

Included in the standard times

a. For contingencies and work other than that performed on individual trees, eg refuelling, day to day maintenance of the saw, clearing brash from racks and rides etc, 15% of the time spent on actual thinning.

b. Personal needs and rest: 22% of the total working time.

N.B. Maintenance to guide bar, mechanical breakdowns in excess of 15 minutes are not included in the allowances and should be paid at time rate.

To be included in the price per standard minute

c. The appropriate incentive allowance.

   Chainsaw usage when thinning and chainsaw snedding is 60%.

5 Method of selecting the standard time

Average tree volume, determined by the tariff system is required to determine the time from the table.

6 Standard times for thinning in Sitka spruce

Average tree cubic metres	Standard minutes per tree	Average tree cubic metres	Standard minutes per tree
0·070	5·3	0·15	8·1
0·075	5·5	0·16	8·4
0·080	5·7	0·17	8·7
0·085	5·9	0·18	8·9
0·090	6·1	0·19	9·2
0·095	6·3	0·20	9·4
0·10	6·5	0·22	9·9
0·11	6·8	0·24	10·4
0·12	7·2	0·26	10·8
0·13	7·5	0·28	11·2
0·14	7·8	0·30	11·6
		0·32	12·0

7 Modifications and variations to the standard times for thinning Sitka spruce

N.B. These modifications and variations are only to be applied when the prevailing conditions or job specification differ from those listed in paragraphs 1 and 2.

a. *Brashing per cent*
   If the brashing percentage is consistently 70% or less
   Add 5%.

b. *Stump treatment*
   Using plastic bottle with brush
   Add 5%.

c. *Ground conditions*
   Where ground conditions are difficult because of rocks, excessive steepness, deep drains etc
   Add up to 10%.

1 Conditions

The standard times apply to stands clear felled under the following conditions:

a. One man working using the methods described in the Appendix.

b. Trees to be subsequently extracted in the pole length, or after partial conversion in the wood into one or two (occasionally three) pieces of sawlogs plus top.

c. Brashing percentage at least 90% of the measurable trees (but see paragraph 7b).

d. Ground conditions average for the species, ie lop and top and ground vegetation not interfering seriously with working, ground reasonably even, slopes not more than 35% (20°). (For more difficult ground conditions see paragraph 7a.)

e. Average tree volume determined by the tariff system (see FC Booklet No 36) or by any other recognised method.

2 Job specification

The standard times apply to the following work:

a. Trees felled in the direction most suitable for extraction or as directed by the supervisor.

b. Pole length working (for marking and measuring sawlogs see paragraph 7d, for marking, measuring and crosscutting during snedding see paragraph 7e).

c. Stumps to be cut as low as possible.

d. When trees are *turned* for the completion of snedding, all branches, brashing and pruning knots should be cut off flush with the stem of the tree. With *other* methods as many branches as possible should be trimmed flush, leaving a few stubs or branches to be trimmed during crosscutting.

e. Tops cut off at a diameter of about 7 cm.

143

f. Tops cut up into lengths not exceeding 1·2 m as required by the supervisor.

g. All cuts, including the felling cut, made as squarely across the tree as possible.

h. Buttresses removed from the butt of the tree after felling.

i. Lop and top to be cleared from roads, major racks and drains as required by the supervisor.

j. Stumps to be treated immediately after felling.

N.B. The time for stump treatment is not included in the standard times—see paragraph 7f.

3 Tools and equipment (including safety equipment)

a. Lightweight anti-vibration chainsaw, of an approved pattern and suitable for chainsaw snedding, together with spares and maintenance tools.

b. Fuel and oil cans.

c. Breaking bar with cant-hook attachment.

d. Plastic or alloy felling wedges.

e. Cant-hook (when large trees are required to be turned for snedding).

f. Logger's harness, logger's tape and pulp hook (when primary conversion or measurement at stump is required).

g. Stump treatment equipment.

h. Protective clothing viz:
Safety helmet BS 2826, 2095 or 5240.
Eye protection.
Ear defenders.
Gloves, incorporating ballistic nylon lining on back of left hand.
Nylon leg guard.
Safety boots, incorporating inner ballistic nylon guard.

4 Allowances

a. The following allowances are included in the standard times:

(i) For contingencies and work other than that performed on individual trees, eg refuelling and day to day maintenance of the saw, clearing lop and top from roads, supervisory visits, etc, 23% of the time spent on clear felling.

N.B. Mechanical breakdowns to the saw in excess of 15 minutes are not included in the allowance and should be paid for separately.

(ii) For personal needs and rest, 25% of the total working time.

b. Payment for worker owned chainsaws should be based on chainsaw utilisation of 65%.

5 Method of selecting the standard time

a. Select the correct table for the working method being used (see Appendix for details) as follows:

   (i) Trees felled on top of previously felled poles which are several ranks deep because felling is *not* immediately followed by extraction. Trees snedded as completely as possible without turning (Working Method 1). See table 6A.

   (ii) Trees felled on a surface largely unobstructed by previously felled poles, for example when extraction follows so closely after felling that normally not more than one rank of previously felled poles (lying parallel) are on the ground when the second rank is being cut.

   Trees turned and completely snedded (Working Method 2a). Trees partially converted at stump, each piece turned and completely snedded (Working Method 2b). See table 6B.

   Trees *not* turned, snedded as completely as possible (Working Method 1). See table 6C.

   Trees *not* turned and deliberately incompletely snedded (Working Method 3). See table 6D.

   N.B. Extra snedding will be required during crosscutting at roadside. See paragraph 7g.

b. Obtain mean tree volume by the tariff system or another recognised method.

c. Select the standard time per tree using the correct table and volume. In cases where it is necessary to subdivide the DBH distribution when tariffing a different mean tree size (and therefore time per tree) should be used for the trees in each subdivision.

Example 1

Trees extracted immediately after felling.
Trees turned at the butt for complete snedding.
An average of $1\frac{1}{2}$ sawlogs of timber per tree marked and measured for subsequent crosscutting.
Moderately difficult terrain.
Mean tree volume $= 0.83 \, m^3$.

Felling and snedding time (from table 6B)	$= 15.73$ SM per tree
Allowance for difficult terrain (from paragraph 7a) 5% of 15.73	$= 0.79$ SM per tree
Allowance for mark and measure (from paragraph 7d) $1\frac{1}{2} \times 0.45$ SM	$= 0.68$ SM per tree
Allowance for stump treatment (from paragraph 7f)	$= \underline{0.80}$ SM per tree
Total time	$= \underline{18.00}$ SM per tree

145

Example 2
Trees extracted immediately after felling.
Trees deliberately incompletely snedded (not turned).
No marking and measuring of sawlogs at stump, crosscutting carried out at roadside by other operators.
Conditions as shown in paragraph 1.
Mean pole volume 0·93 m³.

Felling and snedding time (from table 6D).

Interpolated: $10·97 + \left[\dfrac{(11·41 - 10·97)}{10} \times 3\right]$ = 11·10 SM per tree

Allowance for stump treatment (from 7f) = <u>0·80</u> SM per tree

Total time = <u>11·90</u> SM per tree

N.B. Extra time for snedding should be added to the crosscutting time (see paragraph 7g),

Interpolated $1·08 + \left[\dfrac{(1·17 - 1·08)}{10} \times 3\right]$ = 1·11 SM per tree

6 Standard times for clear felling Douglas fir

A   *Felled trees* **not** *extracted immediately.*
Trees snedded as completely as possible **without** *turning. (Working Method 1).*

Volume of average tree in m³	Standard time per tree in standard minutes
0·3	9·79
0·4	11·35
0·5	12·81
0·6	14·14
0·7	15·34
0·8	16·40
0·9	17·34
1·0	18·25
1·1	19·19
1·2	20·11
1·3	21·02
1·4	21·91
1·5	22·76
1·6	23·60
1·7	24·38
1·8	25·11
1·9	25·77
2·0	26·36

Interpolate as necessary.

B *Felled trees extracted soon after felling.*
   Trees turned and completely snedded or trees partially converted at stump, each piece turned and completely snedded. (Working Method 2a or 2b).

Volume of average tree in m³	Standard time per tree in standard minutes
0·3	10·75
0·4	11·96
0·5	13·06
0·6	14·04
0·7	14·93
0·8	15·73
0·9	16·44
1·0	17·08
1·1	17·65
1·2	18·17
1·3	18·64
1·4	19·07
1·5	19·47
1·6	19·84
1·7	20·20
1·8	20·55
1·9	20·91
2·0	21·28

Interpolate as necessary.

N.B. Where trees are partially converted at stump, extra time should be allowed for marking, measuring and crosscutting the sawlogs, see paragraph 7e.

C *Felled trees extracted soon after felling.*
Trees snedded as completely as possible **without** *turning. (Working Method 1).*

Volume of average tree in m³	Standard time per tree in standard minutes
0·3	8·14
0·4	9·56
0·5	10·84
0·6	11·99
0·7	13·00
0·8	13·90
0·9	14·70
1·0	15·40
1·1	16·01
1·2	16·55
1·3	17·03
1·4	17·45
1·5	17·82
1·6	18·17
1·7	18·49
1·8	18·79
1·9	19·09
2·0	19·40

Interpolate as necessary.

D *Felled trees extracted soon after felling.*
*Trees **not** turned and **incompletely** snedded at stump. (Working Method 3).*

Volume of average tree in m³	Standard time per tree in standard minutes
0·3	7·27
0·4	8·04
0·5	8·74
0·6	9·38
0·7	9·95
0·8	10·48
0·9	10·97
1·0	11·41
1·1	11·82
1·2	12·20
1·3	12·55
1·4	12·89
1·5	13·22
1·6	13·54
1·7	13·86
1·8	14·18
1·9	14·52
2·0	14·87

Interpolate as necessary.

N.B. Extra time will have to be allowed for completing the snedding not done at stump. This time will be *additional to the time for crosscutting at roadside,* see paragraph 7g.

7 Modifications and variations to the standard times
N.B. These modifications and variations are only to be applied when the prevailing conditions or job specification differ from those listed in paragraphs 1 and 2.

a. *Ground conditions*

Where ground conditions make working difficult because of ground broken by rocks or gullies, thick coppice or bramble, or excessive steepness (slopes of more than 35% (20°))

Add up to 10% to the standard times in steps of 5%.

Exceptionally for extremely difficult conditions an addition of 15% to the standard times may be justified.

b. *Brashing*
If the percentage of trees brashed is less than 90%, the brashing of all measurable unbrashed trees should be allowed for.

Add 0·75 standard minutes per tree brashed (or minimal brashing and extra snedding).

c. *Unmeasurable trees and coppice stems*
For felling, crosscutting and disposing of unmeasurable trees or coppice poles if required by the forester

Add 1·0 standard minutes per stem.

d. *Measure and mark sawlog lengths*
When sawlog lengths are measured and marked (but **not** crosscut) during snedding using a logger's tape for measuring and the chainsaw for marking the trees

Add 0·45 standard minutes per tree for each timber length measured and marked.

e. *Measure, mark and crosscut sawlog lengths*
N.B. This modification for use with table 6B **only.**
When sawlog lengths are measured, marked **and** crosscut during snedding, make additions from the following table:

Mean top diameter of sawlog	Time per sawlog in standard minutes
15	0·64
20	0·70
25	0·79
30	0·94
35	1·16

f. *Stump treatment*
For applying fungicide to the stump immediately after felling (including fetching chemicals, filling containers, etc),

Add 0·60 standard minutes per tree for trees up to 0·70 m³,
0·80 standard minutes per tree for trees more than 0·70 m³.

g. *Additional time for snedding*
N.B. **These additions should not be made to the felling times. They are an addition to the time for crosscutting at roadside.**

Some allowance is made in *crosscutting* tables for the trimming of the occasional snags or branches uncut during normal snedding, but with the working method assumed in table 6D a substantial amount of extra snedding may be needed. The time to be allowed for this work is given below. Note that it is not essential for the same operator to do the felling and the crosscutting (with extra snedding).

Volume of average tree in m³	Standard time for extra snedding in standard minutes per tree
0·3	0·84
0·4	0·85
0·5	0·86
0·6	0·89
0·7	0·93
0·8	1·00
0·9	1·08
1·0	1·17
1·1	1·28
1·2	1·39
1·3	1·51
1·4	1·64
1·5	1·77
1·6	1·91
1·7	2·05
1·8	2·19
1·9	2·32
2·0	2·45

Working Method No 1—trees snedded as completely as possible without turning

The normal working method when trees are not turned for snedding (tables 6A and 6C) is as follows:

Walk to tree.
(Clear foot of tree if necessary).
(Chainsaw brash if necessary).
Lay in (and trim buttresses if necessary).
Saw through tree.
Take down.
(Trim stump if necessary).
Treat stump with fungicide.
Trim butt.
Sned up tree, snedding all branches which can be reached from one side of the tree.
Cut off top.
Cut up top as required.
Sned back down tree, removing remainder of branches.

N.B. This method may leave a few stubs or branches to be removed during crosscutting. This work is allowed for in the crosscutting standard times.

Working Method No 2a—trees turned for snedding

The normal working method when trees are turned so that snedding can be completed in the wood (table 6B) is as follows:

Walk to tree.
(Clear foot of tree if necessary).
(Chainsaw brash if necessary).
Lay in (and trim buttresses if necessary).
Saw through tree.
Take down.
(Trim stump if necessary).
Treat stump with fungicide.
Trim butt.
Sned up tree, snedding all the branches easily reached from one side of the tree.
Cut off top.
Cut up top as required.
(Turn tree at top if possible, otherwise:)
Walk to butt.
Turn tree using a cant-hook.
Walk to tip.
Sned back down tree, removing remainder of branches.

Working Method No 2b—trees partly converted, turned and completely snedded at stump

The method is as for method 2a up to 'trim butt', then:

Attach hook of logger's tape to butt of tree.
Sned up tree to top of first sawlog length.
Measure sawlog length.
Mark sawlog length.
Crosscut sawlog length.
(Wind in tape if necessary.)

Repeat for each sawlog length.

Continue as in method 2a, turning and snedding each piece. (When marking and measuring only are done the method is the same except that the trees are not crosscut.) Table 6B as modified by paragraphs 7d and 7e applies.

Working Method No 3—trees not completely snedded in the wood

The normal working method when part of the snedding is left undone until after the tree has been extracted (table 6D) is as follows:

Walk to tree.
(Clear foot of tree if necessary.)
(Chainsaw brash if necessary.)
Lay in (and trim buttresses if necessary).
Saw through tree.
Take down.
(Trim stump if necessary.)
Treat stump with fungicide.
Trim butt.
Sned up tree, snedding all the branches which can be reached from one side of the tree.
Cut off top.
Cut up top as required.

1 Conditions

The standard times apply to stands row thinned under the following conditions:

a. Average tree volume determined by the tariff system (see FC Booklet No 36).

b. The type of thinning is one row in three.

c. The stands have been:
   (i) at least 90% brashed

or (ii) the row to be removed has been wholly brashed as well as the facing side of the row on either side. (See paragraph 7b for unbrashed stands.)

d. Floor conditions are average:
   (i) Flat or slope not more than 5° (9%).
   (ii) Some decayed brash and undergrowth.

2 Job specification

The standard times are for the following work:

a. To fell, sned and where required pile the trees.

b. Stumps to be cut as low as possible.

c. Branches and pruning knots to be cut down to the wood.

d. Tops to be cut off at an average top diameter of 5 cm OB and piled amongst standing trees.

e. Brash to be left in lane created to assist tractor extraction.

f. All cuts including felling cut to be made squarely across tree.

g. Where stacking is required by supervisor, trees to be stacked in piles of not less than 5 and not more than 20 with butts together.

h. Stumps to be treated within 30 minutes of felling; the entire top surface of the stump to be covered.

N.B. The times include making up the suspension and filling the container.

i. See also Appendix.

155

3 **Tools and equipment** (including safety equipment)
  a. Lightweight self-oiling chainsaw of a type suitable for chainsaw snedding, spares and tools.
  b. Steel tape to measure diameters.
  c. Protective clothing viz:
    Safety helmet BS 2826, 2095 or 5240.
    Eye protection.
    Ear defenders.
    Gloves, incorporating ballistic nylon lining on back of left hand.
    Nylon leg guard.
    Safety boots, incorporating inner ballistic nylon guard.
  d. Stump treatment bottle with brush.

4 **Allowances**
  Included in the standard times
  a. For contingencies and work other than that performed on the individual trees, eg refuelling and day to day maintenance of the saw, clearing small trees, clearing tops etc, 20% of the time spent on actual thinning.
  b. Personal needs and rest: 20% of the total working time.
  To be included in the price per standard minute
  c. The appropriate incentive allowance
    Chainsaw usage when thinning and chainsaw snedding is 60%.

5 **Method of selecting the standard time**
  The trees shall be tariffed according to the method specified in FC Booklet No 36. Look up the standard time for the average tree in table 6.

6 **Standard times per tree for line thinning in Corsican pine**

Average tree volume (m³)	ST per tree to fell, sned, pile and treat stump (SMs)
0·01	1·65
0·02	1·91
0·03	2·18
0·04	2·44
0·05	2·77
0·06	2·87
0·07	3·06
0·08	3·25
0·09	3·45

7 Modifications and variations to the standard times

N.B. These modifications and variations are only to be applied when the prevailing conditions or job specification differ from those listed in paragraphs 1 and 2.

a. *Ground conditions*

Where ground conditions are difficult because of the presence of dense bramble etc

Add up to 0·70 SMs per tree.

b. *Unbrashed stands*

Where none of the trees is brashed

Add 20%.

c. *Unmeasurable trees*

For each unmeasurable tree (BHQG of less than 7 cm diameter *allow 0·85 SMs.*

Method of working

The working method to which the standard time table relates is as follows:

(i)   Walk to tree carrying saw (running) and stump treatment bottle.

(ii)   Prepare to fell tree by removing deep litter and occasional brash.

(iii)   Fell tree down rack (ie away from remaining standing trees in row).

(iv)   Take down.

(v)   Trim butt (where necessary).

(vi)   Treat stump.

(vii)   Sned to 5 cm (or other top diameter—as directed).

(viii) Cut off top and put aside to unthinned rows.

(ix)   Turn pole and sned to butt.

(x)   Put aside saw.

(xi)   Stack trees in poles of approximately 5 trees on alternate sides of rack.

(x)   Take up saw and repeat from (i).

Standard Times for Thinning Sitka Spruce Using a Chainsaw
North Wales

1 Conditions

The standard times apply to stands thinned under the following conditions:

a. The average volume per tree determined by 'tariff' system (see FC Booklet No 36).

b. The brashing percentage is at least 80, but see paragraph 7a.

c. The floor conditions are average for the species and locality, ie up to 20° (36%) slope, normal amounts of brash, a few rocks and normal drainage patterns. For more difficult conditions see paragraph 7b.

d. Normal thinning for species and locality.

e. One man working.

2 Job specification

The standard times apply to the following work:

a. Stumps to be cut as low as possible.

b. Branches and pruning knots to be cut off flush with the stem of the tree. Snedding by chainsaw.

c. Lop and top to be cleared from major racks, roads and main drains.

d. Tops to be cut off at a diameter of 7 cm unless otherwise instructed by the supervisor.

e. All cuts including the felling cut, to be made as squarely across the tree as possible.

f. The supervisor will stipulate the direction of extraction and all trees should be felled in a direction to assist the subsequent extraction.

g. All buttressing to be removed from the butt by chainsaw.

159

3 Tools and equipment (including safety equipment)

a. Protective clothing viz:
Safety helmet BS 2826, 2095 or 5240.
Eye protection.
Ear defenders.
Gloves, incorporating ballistic nylon lining on back of left hand.
Nylon leg guard.
Safety boots, incorporating inner ballistic nylon guard.

b. Chainsaw (lightweight) with 30–38 cm bar, file and spares.

c. Fuel cans and gauze funnel.

d. Cant hook, felling tongs or breaking bar.

e. Small alloy wedge.

f. Plastic bottle with brush for stump treatment.

4 Allowances

Included in the standard times

a. For contingencies and work other than that performed on individual trees eg refuelling, day to day maintenance of the saw, clearing brash, general preparation, to and from camp, inspect and consider etc 25% of the time spent thinning and snedding.

b. For personal needs and rest, 22% of the total working time.

N.B. Mechanical breakdowns in excess of 15 minutes should be paid at time rate.

To be included in the price per standard minute

c. The appropriate incentive allowance.

N.B. Chainsaw usage when thinning and snedding is 60%.

5 Method of selecting the standard time

Average tree volume, determined by the tariff system (see FC Booklet No 36), together with the tariff number is required to determine the time from the table in paragraph 6.

6 Standard times for chainsaw thinning and snedding (SMs per tree)
Sitka spruce

Average tree volume (m³)	Tariff No		
	15 – 19	*20 – 24*	*25 – 30*
0·055	6·3	6·1	5·6
0·060	6·5	6·3	5·8
0·065	6·7	6·5	6·0
0·070	6·9	6·7	6·2
0·075	7·1	6·9	6·4
0·080	7·3	7·1	6·6
0·085	7·5	7·3	6·8
0·090	7·7	7·5	7·0
0·095	7·9	7·7	7·2
0·10		7·9	7·4
0·11		8·3	7·8
0·12		8·7	8·2
0·13		9·1	8·6
0·14		9·4	8·9
0·15		9·7	9·2
0·16		10·0	9·5
0·17		10·2	9·7
0·18		10·5	10·0
0·19		10·7	10·2
0·20		11·0	10·5
0·22		11·4	11·0
0·24			11·5
0·26			11·9
0·28			12·3
0·30			12·7
0·32			13·1

7 Modifications and variations to the standard times for chainsaw thinning and snedding Sitka spruce

N.B. These modifications and variations are only to be applied when the prevailing conditions or job specification differ from those listed in paragraphs 1 and 2.

a. *Brashing percentage*
   If the brashing percentage is consistently less than 70% **add** 5%.

b. *Ground conditions*
Where slopes are steeper than 20° (36%), or rocks and coppice are frequent, then **add** up to 12% (in 3% steps).

c. *Take down in first and second thinnings*
Where difficult take down can be attributed to delayed or light thinnings or deep single furrow ploughing **add** up to 10% (in 5% steps).

d. *Putting into tushes*
Making up tushes to facilitate extraction **add** 5%.

e. *Stump treatment*
Standard times for treating stumps (immediately after felling).
    Add 0·40 SMs per tree up to 5 hft (0·18 m³).
        0·60 SMs per tree over 5 hft (0·18 m³).

1 Conditions

The standard times apply to stands clearfelled under the following conditions:

a. The average volume per tree determined by the tariff system (see FC Booklet No 36).

b. All trees brashed.

c. Slopes up to 10° (18%).

d. Sparsely scattered bramble and some coppice.

2 Job specification

The standard times are for the following work:

a. Stumps to be cut as low as possible.

b. Branches and brashing knots to be cut off flush with the stem.

c. Tops to be cut off at a diameter of 7 cm. If otherwise instructed by the supervisor, the tables may not apply and guidance should be sought from the local Work Study Officer.

d. All cuts including the felling cut to be made as squarely across the tree as possible.

e. Trees to be felled in the direction indicated by the supervisor.

f. Lop and top to be cleared from rides, roads, and drains.

g. Buttresses to be trimmed.

3 Tools and equipment (including safety equipment)

a. Lightweight chainsaw of a type suitable for snedding, with spare chain and maintenance tools.

b. Plastic or alloy wedges.

c. Cant hook or breaking bar.

d. Bottle with brush for treating stumps.

e. Protective clothing viz:
Safety helmet BS 2826, 2095 or 5240.
Eye protection.
Ear defenders.
Gloves, incorporating ballistic nylon lining on back of left hand.
Nylon leg guard.
Safety boots, incorporating inner ballistic nylon guard.

4 Allowances

a. For contingencies and work other than that performed on individual trees; 22% of the time spent on felling and snedding for refuelling and day to day maintenance of the saw, clearing brash, counting trees etc.

b. Personal needs and rest 22% of the total working time.
To be included in the price per standard minute

c. The appropriate incentive allowance. In clearfelling of Japanese larch of the volumes shown in paragraph 6, chainsaw usage is 60%.

5 Method of selecting the standard time

a. The trees shall be tariffed according to the method specified in FC Booklet No 36. At the same time the number of whorls of branches to be cut off between butt and the 7 cm top diameter over bark point on the felled sample trees shall be recorded. Where the felled sample tree has been pruned, only the whorls remaining to be cut off shall be counted.

b. To select the standard time from table 6 select the appropriate average volume for the stand (rounding up or down to the nearest) and read off the standard time from col (ii). This time is for the average number of whorls shown in col (iii). If the number of whorls is greater or less than this figure, add/subtract the standard minutes per whorl as shown at the foot of the table. If the average number of whorls is outside the range shown in col (iv), refer to the local Work Study Officer.

6 Standard times per tree for clearfelling Japanese larch

Volume in m³ (i)	Standard time (ii)	Average No of whorls (iii)	Range of whorls (iv)
0·22	4·44	12	9–15
0·24	4·65	13	10–16
0·265	5·00	13	10–16
0·30	5·46	14	11–17
0·325	5·80	15	12–18
0·355	6·20	16	13–19
0·40	6·75	17	14–20
0·425	7·05	18	15–21
0·46	7·48	19	16–22
0·50	7·92	20	17–23

Add/subtract 0·19 SMs per whorl if the average number of whorls is different from the value given in col (iii) but do not go beyond the range given in col (iv).

7 Modifications and variations to the standard times

N.B. These modifications and variations are only to be applied when the prevailing conditions or job specification differ from those listed in paragraphs 1 and 2.

a. *Variations in conditions*

Where the slopes are steeper than 10° (18%) or rocks and coppice and bramble are so frequent that moving about is made difficult, the recommended times in paragraph 6 may be increased by up to 12% in steps of 3%.

b. *Trimming buttresses*

Normal trimming of butts is included in the standard times. Additional work may be required on abnormally rough stems.
    Add 2–4%.

c. *Stump treatment*

Standard time for filling the bottle from ready mixed solution at rideside and applying the solution (the job to be done immediately after felling)
    Add 0·3 SMs per tree up to 0·4 m³.
    Add 0·5 SMs per tree over 0·4 m³ and up to 0·5 m³.

1 Conditions

The standard times apply to stands clearfelled under the following conditions:

a. The volume of the average tree is determined by the tariff system. When in stands previously crown thinned two groups are kept separate (see FC Booklet No 36), a different average tree should be used for each group.

b. The brashing percentage is at least 90% of the measurable trees.

c. Floor conditions are average for the species, ie normal amounts of lop and top from thinnings, little ground vegetation, few rocks, normal drainage patterns. Slopes up to 15° (27%) for areas from which the felled trees are removed during felling (table 6B) and up to 20° (36%) for other areas (table 6A).

d. One man working.

2 Job specification

The standard times apply to the following work:

a. Trees to be felled in the direction most suitable for extraction or as directed by the supervisor.

b. Stumps to be cut as low as possible.

c. Branches and brashing and pruning knots to be cut off flush with the stem of the tree, snedding with the chainsaw. Smaller trees will be turned from the tip for snedding the undersides. Larger trees (above about 0·25 cu m) will normally only be snedded as thoroughly as is possible without turning them (but see paragraph 7g).

N.B. When bigger trees are crosscut at stumps into one or more pieces of timber and a top, it is usual to turn the pieces to sned their undersides after crosscutting. Because it is easier to sned the branches after the pieces have been turned no extra time allowance is normally necessary for the extra turning involved (but see paragraph 7h). The crosscutting should be paid for separately.

d. Lop and top to be cleared from major racks, roads and main drains as required by the supervisor.

e. Tops to be cut off at a diameter of 7 cm.

f. Tops to be cut into lengths not exceeding 1·2 m.

g. All cuts including the felling cut to be made as squarely across the tree as possible.

h. Buttresses to be removed with the chainsaw after felling as required by the supervisor.

i. Stumps to be treated immediately after felling.

N.B. Treating stumps requires extra standard time—see paragraph 7f.

3 **Tools and equipment** (including safety equipment)

a. Lightweight anti-vibration chainsaw of an approved pattern and suitable for chainsaw snedding, together with spares and maintenance tools.

b. Breaking bar which may have a cant-hook attachment.

c. 13 cm plastic felling wedges as required.

d. When larger trees are turned from the butt, or timber pieces are turned after crosscutting, a cant-hook or breaking bar with cant-hook attachment (see 3b).

e. If trees are crosscut at stump, Spencer or Bushman's logger's tape and pulphook.

f. Bottle and brush for stump treatment.

g. Protective clothing viz:
Safety helmet BS 2826, 2095 or 5240.
Eye protection.
Ear defenders.
Gloves, incorporating ballistic nylon lining on back of left hand.
Nylon leg guard.
Safety boots, incorporating inner ballistic nylon guard.

4 **Allowances**

Included in the standard times

a. For contingencies and work other than that performed on individual trees, eg refuelling and day to day maintenance of the saw, clearing lop and top from rides, etc, 23% of the time spent on clearfelling.

N.B. Mechanical breakdowns to the saw in excess of 15 minutes are not included in the allowance and should be paid for at time rates.

b. For personal needs and rest, 25% of the total working time.

To be included in the price per standard minute

c. The appropriate incentive allowance.

d. For calculating the payment for worker-owned chainsaw, chainsaw utilisation is 65% of the total time (including time for stump treatment) when the saw is stopped during stump treatment. If the saw is left running during this operation chainsaw utilisation is 70%. (Utilisation is defined as the time the saw is running plus saw maintenance and refuel times.)

5 Method of selecting the standard time

a. Different felling systems.

   (i) *Felled trees not extracted immediately after felling*

Two quite distinct types of felling are covered by the standard times. In the first of these felling and extraction are two quite separate operations done at different times. Because no extraction is done during felling the cutter is working for most of the time between felled trees, and these and the associated lop and top are a considerable hindrance during snedding. For this type of felling, which is usually on slopes and is often followed by double-drum winch extraction, use table 6A.

   (ii) *Felled trees extracted immediately after felling*

In the second type of felling the cutter is working on comparatively flat ground and trees are skidded out by tractor or otherwise removed soon after felling, so that the cutter is working for most of the time standing on lop and top but without felled trees nearby. For this type of felling use table 6B.

   (iii) *Other types of felling*

Felling is sometimes seen which is intermediate between the two main types, and for such felling the times will fall between the times in the two tables—see paragraph 7a.

b. Trees should be tariffed by the method described in FC Booklet No 36. When the sample trees are measured, the number of whorls of branches cut off between the normal brashing height and the 7 cm diameter mark on the tree should also be recorded. When a felled tree has been pruned, only the whorls remaining unpruned should be counted, and when there are internodal branches growing between the whorls the internodals grown in any year should be included as part of the whorl below. Epicormic shoots on edge trees should be ignored, and whorls below 1·8 m in unbrashed trees should not be included, as brashing with the chainsaw is paid for separately (see paragraph 7c).

c. To select the standard time from table 6A or table 6B select the appropriate average volume or volumes for the stand and read off the standard time(s) from column (ii). The time is for the average number of whorls shown in column (iii). Neither table should be used for stands where the average number of whorls per tree differs by more than 5 from the average given in the tables.

6A Standard times for clearfelling Sitka spruce—felled trees not extracted immediately after felling

Volume of average tree (m^3)	ST per tree in standard minutes	Average number of whorls above brashing height	Volume of average tree (m^3)	ST per tree in standard minutes	Average number of whorls above brashing height
(i)	(ii)	(iii)	(i)	(ii)	(iii)
0·04	4·20	13	0·84	21·02	23
0·06	4·80	13	0·88	21·54	23
0·08	5·40	14	0·92	22·06	23
0·10	5·98	14	0·96	22·56	23
0·12	6·57	14	1·00	23·02	23
0·14	7·12	15	1·04	23·48	23
0·16	7·67	15	1·08	23·92	23
0·18	8·23	16	1·12	24·35	23
0·20	8·75	16	1·16	24·78	23
0·24	9·78	17	1·20	25·18	23
0·28	10·76	18	1·24	25·58	23
0·32	11·72	18	1·28	25·98	23
0·36	12·62	19	1·32	26·37	23
0·40	13·50	19	1·36	26·74	23
0·44	14·33	20	1·40	27·12	23
0·48	15·13	20	1·44	27·49	23
0·52	15·90	21	1·48	27·87	23
0·56	16·64	21	1·52	28·24	23
0·60	17·34	22	1·56	28·61	23
0·64	18·02	22	1·60	29·00	23
0·68	18·67	22	1·64	29·38	23
0·72	19·30	22	1·68	29·77	23
0·76	19·90	23	1·72	30·15	23
0·80	20·46	23	1·76	30·57	23

This table should not be used for stands where the average number of whorls between 1·8 m from the ground and 7 cm diameter on the trunk differs by more than 5 from the figure given in column (iii) above. If times are required for such stands, advice should be sought from the Work Study Branch.

6B Standard times for clearfelling Sitka spruce—felled trees extracted immediately after felling

Volume of average tree (m³) (i)	ST per tree in standard minutes (ii)	Average number of whorls above brashing height (iii)	Volume of average tree (m³) (i)	ST per tree in standard minutes (ii)	Average number of whorls above brashing height (iii)
0·04	3·08	13	0·68	14·05	27
0·06	3·57	13	0·72	14·50	27
0·08	4·06	14	0·76	14·90	27
0·10	4·54	15	0·80	15·30	27
0·12	5·00	16	0·84	15·67	27
0·14	5·44	17	0·88	16·04	27
0·16	5·87	17	0·92	16·39	27
0·18	6·30	18	0·96	16·73	27
0·20	6·72	19	1·00	17·05	26
0·24	7·52	20	1·04	17·37	26
0·28	8·29	22	1·08	17·68	26
0·32	9·01	23	1·12	17·99	26
0·36	9·69	24	1·16	18·30	26
0·40	10·35	24	1·20	18·60	25
0·44	10·96	25	1·24	18·91	25
0·48	11·55	25	1·28	19·22	24
0·52	12·10	26	1·32	19·53	24
0·56	12·64	26	1·36	19·85	24
0·60	13·13	27	1·40	20·17	24
0·64	13·61	27	1·44	20·51	24

This table should not be used for stands where the average number of whorls between 1·8 m from the ground and 7 cm diameter on the trunk differs by more than 5 from the figure given in column (iii) above. If times are required for such stands, advice should be sought from the Work Study Branch.

7 Modifications and variations to the standard times

N.B. These modifications and variations are only to be applied when the prevailing conditions or job specification differ from those listed in paragraphs 1 and 2.

a. *Other types of felling*

Besides felling on to ground where all the previously felled trees are still lying and felling on to ground from which the felled trees have been extracted, there are also felling methods intermediate between these two extremes. For such areas the times in table 6A should be reduced by up to 20% in steps of 5% according to the supervisor's estimate of the snedding difficulty. The times for felling onto areas with no previously felled poles on the ground (table 6B) are approximately 25% below those in table 6A for any given size of tree.

b. *Ground conditions*

Where ground conditions are difficult because of rocks, deep drains or excessive steepness (slopes of more than 20° 36%) add up to 10% in 5% steps. For extremely difficult conditions add 15%.

c. *Brashing percentage*

If the percentage of trees brashed is less than 90% the brashing of all measurable unbrashed trees should be paid for. The time for chainsaw brashing (or minimal brashing and extra snedding) is 0·75 SM per tree.

d. *Unmeasurable trees*

If unmeasurable trees are felled and cut up into 1·2 m lengths the standard time for this operation is 1·00 SM per stem.

N.B. Extra brashing and the disposal of unmeasurable trees may be paid for separately from the felling of the crop. Alternatively, the total time for brashing and/or the felling and disposal of unmeasurable trees may be worked out and divided by the number of the crop trees, the resulting figure being added to the standard time per tree, and this method is to be preferred.

e. *Moving tips to racks, disposing of lop and top*

For moving the tips of felled trees to an extraction rack and clearing lop and top from the rack when this is done in clear-felling areas to facilitate extraction, add 0·10 SMs per tree, to be added to all trees.

f. *Stump treatment*

For applying nitrite, polybor or urea to the stump immediately after felling (including fetching chemicals, filling cans, etc) add 0·40 SMs per tree for trees up to 0·18 cu m and 0·60 SMs per tree for trees from 0·18 to 0·72 cu m and 0·80 SMs per tree for trees over 0·72 cu m in volume.

g. *Turning trees for snedding*

The normal method of snedding trees which are too large to be turned from the top is to sned them as thoroughly as possible without turning them. When it is considered necessary to turn such trees from the butt using a cant-hook add 3% to the felling standard times.

h. *Turning timber for snedding*

When trees are crosscut at stump into one or more pieces of timber and a pole it is usual to turn the pieces during snedding. Normally this work is covered by the standard times, but in difficult conditions or when trees with a volume of more than

0·75 cum or pieces of more than 0·25 cum are being crosscut an addition to the felling times may be necessary. When in such conditions it is considered essential to turn the timber to ensure an adequate standard of snedding, *add* 5% to the felling standard times.

1 Conditions

The standard times apply to stands thinned under the following conditions:

a. The average volume per tree is determined by 'tariff' system (see FC Booklet No 36).

b. The brashing percentage is at least 80.

c. The floor conditions are average for species and locality, ie slopes of up to 15° (27%), normal amounts of brash and a few rocks.

d. Normal thinning for species and locality.

e. One man working.

2 Job specification

The standard times apply to the following work:

a. Stumps to be cut as low as possible.

b. Branches, brashing and pruning knots to be cut off flush with the stem of the tree. Snedding by chainsaw. Smaller trees to be turned from the tip for snedding the undersides. Larger trees (above about 0·25 cu m) to normally be snedded as thoroughly as possible without turning them.

c. Tops to be cut off at a diameter of 6–7 cm unless otherwise instructed by the supervisor.

d. All cuts including the felling cut to be made as squarely across the tree as possible.

e. Trees to be felled in the direction indicated by the supervisor.

f. All buttressing to be removed from the butt as required by the supervisor.

g. Lop and top to be cleared from major roads.

3 **Tools and equipment** (including safety equipment)

   a. Chainsaw (lightweight) of an approved pattern with 30–38 cm bar, file and spares.

   b. Fuel can and gauze funnel.

   c. Breaking bar.

   d. Suitable equipment for stump treatment.

   e. Protective clothing viz:
Safety helmet BS 2826, 2095 or 5240.
Eye protection.
Ear defenders.
Gloves, incorporating ballistic nylon lining on back of left hand.
Nylon leg guard.
Safety boots, incorporating inner ballistic nylon guard.

4 **Allowances**

Included in the standard times

   a. For contingencies and work other than that performed on individual trees eg day to day maintenance of the saw, clearing brash, general preparation, etc 25% of the time spent on felling and snedding.

   b. For personal needs and rest 24% of the total working time.

To be included in the price per standard minute

   c. The appropriate incentive allowance.

   d. For calculating the payment for worker-owned saws, chainsaw utilisation is 70% of the total working time.

5 **Method of selecting the standard time**

   a. During tariffing the number of whorls of branches to be cut off between the butt and 7 cm top diameter over bark on the felled sample trees shall be recorded. Where the felled sample tree has been pruned, only the whorls remaining to be cut off should be recorded.

   b. To select the standard time from table 6, select the appropriate average volume for the stand and read off the standard time from col (ii). This time is for the average number of whorls shown in col (iii). If the number of whorls differs from this figure, add/subtract the standard minutes per whorl as shown at the foot of the table. If the average number of whorls varies by more than + or − 4 whorls from col (iii) refer to the Work Study Officer.

N.B. For the purposes of this table a whorl consists of 3 or more branches or parts of branches originating from a common node. (The variations may be due to natural pruning or breakages after felling.) Ignore internodal branches and count multiple whorls as one unless there is at least 30 centimetres between separate whorls in the multiple whorls.

6 Standard time per tree for chainsaw thinning and snedding in Japanese larch

Volume of average tree in m³ (i)	Standard time per tree in standard minutes (ii)	Average No of whorls per tree (iii)
0·05	3·5	7·6
0·08	4·2	9·0
0·11	4·8	10·4
0·14	5·4	11·5
0·17	5·9	12·6
0·20	6·4	13·6
0·23	6·8	14·4
0·26	7·2	15·1
0·29	7·6	15·8
0·32	7·9	16·3
0·35	8·2	16·7
0·38	8·5	17·1
0·41	8·7	17·3
0·44	9·0	17·5
0·47	9·2	17·6

Add/subtract 0·2 standard minutes per whorl if the average number of whorls is different from the value given in col (iii) up to a limit of + or − 4 whorls. For stands where the difference is more than 4 whorls consult Work Study Officer.

7 Modifications and variations to the standard times

N.B. These modifications and variations are only to be applied when the prevailing conditions or job specification differ from those listed in paragraphs 1 and 2.

a. *Ground conditions*
Where slopes are steeper than 15° (27%), or rocks and coppice are frequent,
Add up to 15% in steps of 5%.

b. *Take down in first and second thinnings*
Where difficult take down can be attributed to delayed or light thinnings,
Add up to 10%.

c. *Stump treatment*
Standard times for applying nitrite to the stump immediately after felling
Add 0·40 SMs per tree up to 0·18 m³.
0·60 SMs per tree over 0·18 m³.

Standard Times for Clearfelling European Larch Using a Chainsaw
East Scotland

1 Conditions

The standard times apply to stands clearfelled under the following conditions:

a. Tree volume assessment by tariff system (see FC Booklet No 36) is based on measurement to 7 cm top diameter.

b. The brashing percentage is at least 90.

c. Slopes of up to and including 15° (27%).

d. Sparsely scattered bramble, raspberry and broom.

e. Partially decayed lop and top and stumps present from earlier thinnings.

f. Some rocks present.

g. One man working.

2 Job specification

The standard times apply to the following work:

a. Stumps to be cut as low as possible.

b. Branches and brashing knots to be cut flush with the stem.

c. Tops to be cut off at a diameter of 6–7 cm.

d. Mill timber to be cut off at 18 cm top diameter.

e. All cuts, including the felling cut to be made as squarely across the tree as possible.

f. Trees to be felled in the direction most suitable for extraction or as directed by the supervisor.

g. Lop and top to be cleared from rides, roads and drains.

h. Buttresses to be removed with chainsaw after felling as required by the supervisor.

i. Stumps to be treated immediately after felling.

N.B. Treating stumps requires extra standard time—see paragraph 7b.

3 Tools and equipment (including safety equipment)

a. Lightweight AV chainsaw of a type suitable for snedding, with spare chain and maintenance tools.

b. Plastic or alloy wedges.

c. Cant hook or breaking bar.

d. Suitable equipment for treating stumps.

e. Protective clothing viz:
Safety helmet BS 2826, 2095 or 5240.
Eye protection.
Ear defenders.
Gloves, incorporating ballistic nylon lining on back of left hand.
Nylon leg guard.
Safety boots, incorporating inner ballistic nylon guard.

4 Allowances

Included in the standard times

a. For contingencies and work other than that performed on individual trees, eg refuelling and day to day maintenance of the saw, clearing lop and top from rides etc, 26% of the time spent on clearfelling.

b. Personal needs and rest, 23% of the total working time.

To be included in the price per standard minute

c. The appropriate incentive allowance.

5 Method of selecting the standard time

Average tree volume by tariff (see FC Booklet No 36) is required to determine the time from the table in paragraph 6.

6 Standard time per tree for clearfelling European larch (chainsaw snedding)

Average tree volume in m³	Standard time per tree	Average tree volume in m³	Standard time per tree
0·16	4·7	0·36	7·2
0·17	4·9	0·38	7·4
0·18	5·0	0·40	7·6
0·19	5·2	0·45	8·1
0·20	5·4	0·50	8·5
0·22	5·6	0·55	8·9
0·24	5·8	0·60	9·2
0·26	6·1	0·65	9·5
0·28	6·3	0·70	9·9
0·30	6·6	0·75	10·2
0·32	6·8	0·80	10·5
0·34	7·0		

7 Modifications and variations to the standard times

N.B. These modifications and variations are only to be applied when the prevailing conditions or job specification differ from those listed in paragraphs 1 and 2.

a. *Ground conditions*

1 Where the slope is more than 15° (27%) and up to 25° (47%)
Add 5%.

2 Where the slope is steeper than 25° (47%)
Add up to 10%.

3 Where work is impeded by the presence of lop, top and stumps from earlier thinnings and the terrain is moderately rough
Add up to 5%.

4 Where the terrain is very broken with rock outcrops or boulders and heavy ground vegetation seriously hinders work
Add up to 20% in steps of 5%.

b. *Stump treatment*

Standard time for filling can or bottle from barrel and applying to stumps immediately after felling
Add 0·7 SMs per tree.

1 Conditions

The standard times apply to stands clearfelled under the following conditions:

a. The volume of the average tree is determined by the tariff system. When in stands previously crown thinned it is necessary to subdivide the DBH distribution (see FC Booklet No 36 paragraph 26 (ix) page 12); a different average tree size should be used for each sub-division when calculating the standard times.

b. The brashing percentage is at least 90% of the measurable trees (see paragraph 7c).

c. Floor conditions are average for the species, ie normal amounts of lop and top from thinnings, little ground vegetation, few rocks, normal drainage patterns. Slopes up to 25% (14°) for areas from which the felled trees are removed during felling (table 6B) and up to 35% (19°) for other areas (table 6A).

d. One man working.

2 Job specification

The standard times apply to the following work:

a. Trees to be felled in the direction most suitable for extraction or as directed by the supervisor.

b. Stumps to be cut as low as possible.

c. Branches and brashing and pruning knots to be cut off flush with the stem of the tree, snedding with the chainsaw. Smaller trees will be turned from the tip for snedding the undersides. Larger trees (above about 0·25 cu m) will normally only be snedded as thoroughly as is possible without turning them (but see paragraph 7g).

N.B. When bigger trees are crosscut at stump into one or more pieces of timber and a top, it is usual to turn the pieces to sned their undersides after crosscutting. Because it is easier to sned the branches after the pieces have been turned no extra time allowance is normally necessary for the extra turning involved (but see paragraph 7h). The crosscutting should be paid for separately.

    d. Lop and top to be cleared from major racks, roads and main drains as required by the supervisor.

    e. Tops to be cut off at a diameter of 7 cm.

    f. Tops to be cut into lengths not exceeding 1·2 m.

    g. All cuts including the felling cut to be made as squarely across the tree as possible.

    h. Buttresses to be removed with the chainsaw after felling.

    i. Stumps to be treated immediately after felling.

N.B. Treating stumps requires extra standard time—see paragraph 7f.

3. Tools and equipment (including safety equipment)

    a. Lightweight anti-vibration chainsaw of an approved pattern and suitable for chainsaw snedding, together with spares and maintenance tools.

    b. Breaking bar which may have a cant-hook attachment.

    c. 13 cm plastic felling wedges as required.

    d. When larger trees are turned from the butt, or timber pieces are turned after crosscutting, a cant-hook or breaking bar with cant-hook attachment (see 3b).

    e. If trees are crosscut at stump, Spencer or Bushman's logger's tape and pulphook.

    f. Suitable equipment for stump treatment.

    g. Protective clothing viz:
Safety helmet BS 2826, 2095 or 5240.
Eye protection.
Ear defenders.
Gloves, incorporating ballistic nylon lining on back of left hand.
Nylon leg guard.
Safety boots, incorporating inner ballistic nylon guard.

4 Allowances

Included in the standard times

    a. For contingencies and work other than that performed on individual trees, eg refuelling and day to day maintenance of the saw, clearing lop and top from rides, etc, 23% of the time spent on clearfelling.

N.B. Mechanical breakdowns to the saw in excess of 15 minutes are not included in the allowance and should be paid for at time rates.

b. For personal needs and rest, 25% of the total working time.

To be included in the price per standard minute

c. The appropriate incentive allowance.

d. For calculating the payment for worker-owned chainsaws, chainsaw utilisation is 70% of the total working time when the chainsaw is running during stump treatment or when stumps are not treated. When the saw is not running during stump treatment chainsaw utilisation is 70% of the standard time for felling or about 65% of the total working time.

5 Method of selecting the standard time

a. *Different working systems*

Two quite distinct types of harvesting operations are covered by the standard times: other types may be thought of as intermediate.

(i)   Felled trees *not* extracted immediately after felling. When felling and extraction are two separate operations done at different times the feller is working for most of the time between felled trees: and these and the lop and top caught between them are a considerable hindrance during snedding. This type of working often occurs on slopes where it is followed by double-drum winch extraction but may occur elsewhere. For this type of working use table 6A.

(ii) Felled trees extracted immediately after felling. This usually occurs when the feller is working along a face or on more than one short face and extraction is close enough behind him for the ground to be cleared of felled trees before he starts along the face again; but it may also occur with other systems. For methods where the cutter is working for most of the time standing on lop and top but on ground cleared of previously-felled trees use table 6B.

(iii) Intermediate between (i) and (ii).

Some felling methods are intermediate between the two main types. For such felling see paragraph 7a.

b. Trees should be tariffed by the method described in Forestry Commission Booklet No 36.

c. To select the standard time from table 6A or table 6B select the appropriate average volume or volumes for the stand and read off the standard time(s) from col (ii).

6A Standard times for clearfelling Norway spruce—felled trees not extracted immediately after felling

Volume of average tree in m^3	Standard time per tree in minutes	Volume of average tree in m^3	Standard time per tree in minutes
0·04	4·17	0·72	17·32
0·06	4·74	0·76	17·77
0·08	5·29	0·80	18·19
0·10	5·83	0·84	18·59
0·12	6·36	0·88	18·96
0·14	6·87	0·92	19·31
0·16	7·38	0·96	19·64
0·18	7·87	1·00	19·95
0·20	8·35	1·04	20·23
0·24	9·28	1·08	20·50
0·28	10·16	1·12	20·75
0·32	11·00	1·16	20·99
0·36	11·80	1·20	21·21
0·40	12·56	1·24	21·41
0·44	13·27	1·28	21·61
0·48	13·95	1·32	21·79
0·52	14·60	1·36	21·96
0·56	15·21	1·40	22·13
0·60	15·78	1·44	22·29
0·64	16·33	1·48	22·44
0·68	16·84	1·50	22·52

6B Standard times for clearfelling Norway spruce—felled trees extracted immediately after felling

Volume of average tree in m^3	Standard time per tree in minutes	Volume of average tree in m^3	Standard time per tree in minutes
0·04	2·69	0·52	11·90
0·06	3·23	0·56	12·37
0·08	3·75	0·60	12·80
0·10	4·26	0·64	13·19
0·12	4·76	0·68	13·55
0·14	5·24	0·72	13·88
0·16	5·70	0·76	14·17
0·18	6·15	0·80	14·44
0·20	6·59	0·84	14·68
0·24	7·43	0·88	14·90

Volume of average tree in m³	Standard time per tree in minutes	Volume of average tree in m³	Standard time per tree in minutes
0·28	8·21	0·92	15·10
0·32	8·94	0·96	15·27
0·36	9·62	1·00	15·42
0·40	10·26	1·04	15·56
0·44	10·85	1·08	15·68
0·48	11·40	1·10	15·74

7 Modifications and variations to the standard times

N.B. These modifications and variations are only to be applied when the prevailing conditions or job specification differ from those listed in paragraphs 1 and 2.

a. *Other types of felling*

Besides felling on to ground where all the previously felled trees are still lying and felling on to ground from which the felled trees have been extracted, there are also felling methods intermediate between these two extremes. For such areas the times in table 6A should be reduced by up to 20% in steps of 5% according to the supervisor's estimate of the snedding difficulty. The times for felling onto areas with no previously felled poles on the ground (table 6B) are approximately 20% below those in table 6A for trees larger than 0·2 m³, and more than 20% less for smaller trees.

b. *Ground conditions*

Where ground conditions are difficult because of rocks, deep drains or excessive steepness (eg slopes in excess of 35% (19°)) add up to 10% to the standard times in steps of 5%. *Exceptionally* for extremely difficult conditions, an additional 15% may be justified.

c. *Brashing percentage*

If the percentage of trees brashed is less than 90% the brashing of all measurable unbrashed trees should be paid for. The time for chainsaw brashing (or minimal brashing and extra snedding) is 0·75 SM per tree.

d. *Unmeasurable trees*

If unmeasurable trees are felled and cut up into 1·2 m lengths the standard time for this operation is 1·00 SM per stem.

N.B. Extra brashing and the disposal of unmeasurable trees may be paid for separately from the felling of the crop. Alternatively, the total time for brashing and/or the felling and disposal of unmeasurable trees may be worked out and divided by the number of the crop trees, the resulting figure being added to the standard time per tree, and this method is to be preferred.

e. *Moving tips to racks, disposing of lop and top*
For moving the tips of felled trees to an extraction rack and clearing lop and top from the rack when this is done in clear-felling areas to facilitate extraction, add 0·10 SMs per tree to all trees.

f. *Stump treatment*
For applying nitrite, polybor or urea to the stump immediately after felling (including fetching chemicals, filling cans, etc) add 0·40 SMs per tree for trees up to 0·18 cu metres.
0·60 SMs per tree for trees from 0·18 to 0·72 cu metres.
0·80 SMs per tree for trees over 0·72 cu metres.

g. *Turning trees for snedding*
The normal method of snedding trees which are too large to be turned from the top is to sned them as thoroughly as is possible without turning them. When it is considered necessary to turn such trees from the butt using a cant-hook add 5% to the felling standard times.

h. *Turning timber for snedding*
When trees are crosscut at stump into one or more pieces of timber and a pole it is usual to turn the pieces during snedding. Normally this work is covered by the standard times, but in difficult conditions or when trees with a volume of more than 0·75 cu m or pieces of more than 0·25 cu m are being crosscut an addition to the felling times may be necessary. When in such conditions it is considered essential to turn the **timber** to ensure an adequate standard of snedding, **add 5%** to the felling standard times.

i. *Trees snedded to top diameters less than 7 cm*
When cutters are required to sned poles beyond 7 cm top diameter additions should be made to the standard times as shown in the table below. The extra time for snedding to the average cut-off diameter should be added to all trees.

Snedding from 7 cm top diameter to	*Standard time per tree in minutes*
6 cm top diameter	0·31
5 cm top diameter	0·58
4 cm top diameter	0·84
3 cm top diameter	1·08

N.B. These times should not be used for trees averaging more than 0·20 m³ in volume to 7 cm.

Standard Times for
Thinning Norway Spruce
Selective and Systematic Thinning

N.B. Systematic thinning includes row, chevron and staggered chevron thinning.

1. Conditions

The standard times apply to crops thinned under the following conditions:

a. Selective or systematic thinning in brashed crops, and systematic thinning in unbrashed crops. For selective thinning in unbrashed crops see paragraph 7b.

b. Brashing percentage of brashed crops at least 70%.

c. Ground conditions normal for the species, viz: up to 30% (17°) slope, few rocks and normal drainage patterns. For more difficult conditions see paragraph 7c.

d. Racks for systematic thinning may be one or two rows wide, or an equivalent width across the rows.

e. Trees readily identified (eg in rows, or marked) before thinning. If the operator has to align side racks during thinning this work should be allowed for separately, see paragraph 7d.

f. Take down problems normal. For difficult take down see paragraph 7e.

g. One man working using a chainsaw for felling and snedding.

h. Working method as described in the Appendix.

i. Mean tree volume to be determined by tariffing (see Booklet No 36) or by any other recognised method.

2 Job specification

The standard times are for the following work:

a. Pole length working. For making and measuring sawlogs at stump see paragraph 7k. For measuring and crosscutting sawlogs at stump see paragraph 7l.

b. Trees to be felled for tip first extraction unless otherwise stipulated by the supervisor. For uphill felling see paragraph 7e.

c. Stumps to be cut as low as possible.

d. Cuts, including the felling cut, to be made squarely across the pole.

e. Branches and pruning knots to be cut off flush with the stem of the trees. Smaller trees (less than 0·25 m³) to be turned at the tip. Large trees not normally turned, but snedded as thoroughly as possible without turning, but see paragraph 7g.

f. All buttressing to be removed from the butts.

g. Tops to be cut off at a diameter of about 7 cm. If snedding is required to a smaller top diameter the extra work should be allowed for, see paragraph 7f.

h. In systematic thinning tops and some branches to be disposed of to the side of or out of the rack. If almost complete removal of branches is necessary the extra work should be allowed for, see paragraph 7h.

i. Lop and top to be cleared from roads and main drains.

j. Stumps to be treated immediately after felling. This work should be allowed for separately, see paragraph 7i.

k. Where stacking of smaller poles ready for chokering is required the work should be allowed for separately, see paragraph 7j.

l. Where felling of unmeasurable trees is required by the supervisor this work should be allowed for, see paragraph 7a.

3 **Tools and equipment** (including safety equipment)

a. A lightweight anti-vibration chainsaw of an approved type, suitable for chainsaw snedding, and tool kit for chainsaw maintenance.

b. Fuel and oil cans.

c. Breaking bar with cant-hook attachment.

d. Alloy or plastic felling wedges.

e. Stump treatment equipment.

f. Protective clothing viz:
Safety helmet BS 2826, 2095 or 5240.
Eye protection.
Ear defenders.
Gloves, incorporating ballistic nylon lining on back of left hand.
Nylon leg guard.
Safety boots, incorporating inner ballistic nylon guard.

g. Logger's harness, spring loaded logger's tape and a pulp hook where primary conversion or measurement at stump is required.

4 **Allowances**

a. The following allowances are included in the standard times:

(i) For contingencies and work other than that performed on

individual trees, eg refuelling and day to day maintenance of the saw, clearing lop and top from roads and drains, fetching tools and supplies etc, 28% of the time actually spent felling.

N.B. Mechanical breakdowns to the saw in excess of 15 minutes are not included in the allowance and should be allowed for separately.

    (ii) For personal needs and rest 25% of the total working time.

b. Payment for worker owned chainsaws should be based on chainsaw utilisation of 70% of the total time.

5   Method of using the Output Guide

a. Select the table appropriate to the thinning method as follows:
Unbrashed crops, systematic thinning—table 6A.
Brashed and partially brashed crops, selective and systematic thinning—table 6B.

b. Obtain mean tree volume by tariffing (see Booklet No 36) or by any other recognised method.

c. Read the time per tree from the table, and modify as necessary with reference to paragraph 7.

Example 1

Selective thinning in crop 70% brashed.
Mean tree volume 0·08 m³.
Conditions etc as in paragraph 1 and 2 of this STT.

Time per tree (from table 6B)	5·01 SMs per tree
Allowance for stump treatment (from paragraph 7i)	0·30 SMs per tree
Total time	5·31 SMs per tree

Example 2

Chevron thinning in unbrashed crop. Ten trees per side rack (not marked).
Difficult take down due to uphill felling.
50% of trees turned at the butt.
Mean tree volume 0·14 m³.
Other conditions etc as in paragraph 1 and 2 of this STT.

Time per tree (from table 6A)	7·67 SMs per tree
Allowance for marking side racks 1 SM per 10 trees (from paragraph 7d)	0·10 SMs per tree
Allowance for uphill felling (from paragraph 7e)	0·20 SMs per tree
Allowance for turning at the butt (from paragraph 7g) 50% of 0·65 SMs	0·33 SMs per tree
Allowance for stump treatment (from paragraph 7i)	0·40 SMs per tree
Total time	8·70 SMs per tree

Example 3

Selective thinning in crop 80% brashed.

Mean tree volume 0·79 m³.

An average of 1½ sawlogs of 25 cm top diameter measured and crosscut per tree, at stump.

Sawlogs and remainder of the pole completely snedded.

Other conditions etc as in paragraph 1 and 2 of this STT.

Time per tree (from table 6B)

interpolated $16·49 + \left[\dfrac{(16·83 - 16·49)}{4} \times 3 \right]$ 16·745 SMs per tree

Allowance for measuring, marking and crosscutting sawlogs (from paragraph 7l(i)) 1½ × 0·79 SMs   1·185 SMs per tree

Allowance for turning tree (or sawlogs) (from paragraph 7l(ii))   0·90   SMs per tree

Allowance for stump treatment (from paragraph 7i)   0·70   SMs per tree

            Total time   19·530 SMs per tree

6A   Standard times for systematic thinning in unbrashed crops

Mean tree volume (m³)	Standard time per tree in standard minutes
0·04	4·61
0·05	4·94
0·06	5·27
0·07	5·59
0·08	5·90
0·09	6·21
0·10	6·51
0·11	6·81
0·12	7·10
0·13	7·39
0·14	7·67
0·15	7·94
0·16	8·21
0·17	8·47
0·18	8·73
0·19	8·98
0·20	9·23

N.B. Interpolate as necessary.

6B Standard times for systematic and selective thinning in brashed and partly brashed crops

Mean tree volume (m³)	Standard time per tree in standard minutes
0·04	3·71
0·05	4·05
0·06	4·37
0·07	4·69
0·08	5·01
0·09	5·32
0·10	5·62
0·11	5·91
0·12	6·21
0·13	6·49
0·14	6·77
0·15	7·04
0·16	7·31
0·17	7·58
0·18	7·83
0·19	8·09
0·20	8·33
0·24	9·28
0·28	10·14
0·32	10·94
0·36	11·67
0·40	12·34
0·44	12·96
0·48	13·52
0·52	14·04
0·56	14·52
0·60	14·97
0·64	15·38
0·68	15·77
0·72	16·13
0·76	16·49
0·80	16·83
0·84	17·16

N.B. Interpolate as necessary.

7 Modifications and variations to the times in the tables

N.B. These modifications and variations are only to be applied when the prevailing conditions or job specification differ from those listed in paragraphs 1 and 2.

a. *Unmeasurable trees*

The time taken to dispose of unmeasurable trees is not included in the times in table 6A and table 6B.

When unmeasurable trees are felled, cut up if necessary, and disposed of at the side of the rack or in adjoining rows, the time for this operation is:

(i) in unbrashed stands 0·9 SMs per tree;
(ii) in brashed stands 0·8 SMs per tree.

When unmeasurable trees are felled and snedded to 3 or 4 cm top diameter, the time for the operation is:

(i) in unbrashed stands 4·0 SMs per tree;
(ii) in brashed stands 3·6 SMs per tree.

N.B. Unmeasurable trees usually need brashing whether in brashed or in unbrashed stands. The time for snedded trees includes brashing but not stacking the pole, see paragraph 7j.

b. *Chainsaw brashing in unbrashed stands*

Times for brashing are included in table 6A for systematic thinning, but where unbrashed crops are receiving a second or subsequent thinning (which is selective)

Add 0·65 SMs per tree to the times in table 6B.

c. *Difficult terrain*

Where ground conditions are difficult because of excessive steepness, deep drains, exceptionally rocky ground etc

Add up to 10% (in 5% steps).

d. *Positioning side racks in chevron thinning*

If side racks have not been marked out in advance and this is done by the operator during thinning

Add one standard minute per side rack.

e. *Difficult take down*

(i) in single row thinning, or in side racks of similar width, in badly delayed thinning or in very closely planted or uneven crops

Add up to 5% of the felling time;

(ii) for uphill felling

Add 0·2 standard minutes per tree.

f. *Cutting to top diameter of less than 7 cm*

If the operator is *requested* to sned trees to top diameter smaller than 7 cm the extra snedding should be allowed for as follows:

Top diameter	Addition per tree
3	0·90 SMs
4	0·70 SMs
5	0·50 SMs
6	0·30 SMs

g. *Turning trees at the butt for completion of snedding*

It is normal for a small proportion (10% approx) of the trees to be turned at the butt using a breaking bar/cant-hook, and this is allowed for in the times. Trees larger than 0·25 m³ will normally be snedded as completely as possible without turning. On difficult areas, however, it may be necessary to turn up to half the trees at the butt, or if sawlogs are measured and crosscut at stump, each log may have to be turned so that it, and the remaining pole, may be completely snedded. This work should be allowed for as shown in the following table:

Mean tree volume m³	Additional time per tree turned
Up to 0·15	0·65 SMs
0·151–0·25	0·70 SMs
0·251–0·50	0·80 SMs
0·51 –0·85	0·90 SMs

h. *Clearing branches from racks*

If the operator is *requested* to move almost all of the branches out of the rack during felling, this extra work should be allowed for

Add 0·10 standard minutes per tree.

i. *Stump treatment*

The following additions should be made for applying a fungicide to the stumps immediately after felling:

Mean tree volume m³	Time per tree for stump treatment
Less than 0·10	0·30 SMs
0·10–0·149	0·40 SMs
0·15–0·199	0·50 SMs
0·20–0·699	0·60 SMs
More than 0·70	0·70 SMs

j. *Stacking poles*

Stacking the snedded poles into twos, threes and fours with their tips together for chokering is standard practice in many forests but has not been included in the standard times.

It is usual to move one or more small poles to a larger pole which is not moved and the number of poles moved to stacks for chokering may be obtained by deducting the number of stacks from the number of poles.

For each pole moved

Add 0·25 standard minutes.

195

k. *Measure and mark sawlog lengths*

When sawlog lengths are measured and marked, but **not** crosscut, during snedding using a logger's tape for measuring and the chainsaw for marking the trees

Add 0·45 standard minutes per tree for each timber length required.

l. *Measure and crosscut sawlogs at stump*

Where sawlogs are measured, marked and crosscut during snedding make the following additions:

(i) the time for actually measuring, marking if necessary, and crosscutting is as follows interpolating as necessary;

Mean top diameter of sawlog (cm)	Additional time per sawlog (SMs)
15	0·64
20	0·70
25	0·79
30	0·94
35	1·16

(ii) *occasionally,* where sawlogs cannot be completely snedded without turning

Add the appropriate time per tree for turning from paragraph 7g.

Working method

The normal working method to which the times in the tables apply is as follows:

(i)   Walk to tree.

(ii) Chainsaw brash, unbrashed or partially brashed trees up to a maximum of shoulder height (see Note (a)).

(iii) Clear any obstruction, undergrowth or debris from around base of tree.

(iv) Decide on felling direction and cut out the 'sink'.

(v)   Saw through.

(vi) Take down using the appropriate aid tool.

(vii) Treat stump.

(viii) Trim butt (when necessary).

(ix) Sned, working up the tree, cutting off all branches which can be reached from one side of the tree.

(x)   Cut off top (and cut up top when necessary).

(xi) Dispose of top at side of rack or in convenient place.

(xii) Turn tree and if tree is not to be stacked position tip ready for extraction (see Note (b)).

(xiii) Finish snedding, working back down the tree.

(xiv) Stack pole, if small enough, ready for chokering (see Note (c)).

Notes:

(a) It is possible to fell after brashing just enough branches to allow access to the tree. The method described above does not increase the total time for the tree.

(b) Small trees are turned at the tip by hand, or by using a branch left on for the purpose. Turning trees larger than $0.25m^3$ is not recommended though it is occasionally necessary, in which case turning should be done at the butt using a breaking bar/cant hook. Extra time should be allowed for this work.

(c) Stacking for chokering is usually done by moving smaller poles to bigger ones. Tips of bigger trees may sometimes be brought level ready for chokering by varying the cut-off diameter.

(d) In chevron thinning it might be easier to fell and extract the main rack trees before felling the side rack trees, but this will depend on the extraction method and the presentation of poles.

Clearance of Diseased Elm Trees

---

1 Conditions

This guide applies to elm trees felled under the following conditions:

a. Mean tree volume not exceeding 1·5 m³.

b. Trees growing in a hedgerow.

c. Impediments associated with hedgerow growth ie felling in one direction, the presence of a fence and ditch, associated hedgerow vegetation.

d. Supplies of chainsaw fuel and oil to be readily available.

2 Job description

This guide covers the following work:

a. A team of 3 men.

b. Cutting fence and turning back as necessary.

c. Securing winch rope using a ladder if necessary.

d. Manoeuvring Land Rover as necessary for winching.

e. Lay in and fell tree using chainsaw.

f. Trim out using chainsaw.

g. Crosscut tree at 20–24 cm top diameter.

h. Cut up and burn crown, branchwood and cordwood.

i. Peel partially using chainsaw.

j. Replace fence.

k. Clear and burn unmeasurable trees and hedging (unmeasurable trees are those containing under 0·1 m³ of usable timber).

3 Tools and equipment (including safety equipment)

a. Three lightweight self oiling chainsaws with 38 cm guide bars and spares and maintenance tools.

b. Land Rover fitted with Mayflower winch at front or a separate Tirfor T16 winch.

c. Trewhella ground anchors.

d. Snatch block.

e. Alloy ladder.

f. Felling wedges and sledgehammers.

g. Cant-hook.

h. Slasher and axe for clearing undergrowth.

i. Wire cutting tools.

j. Tools, stakes and barbed wire for replacing fencing.

k. Old tyres and shovel for fire making.

l. Protective clothing viz:
Safety helmet BS 2826, 2095 or 5240.
Eye protection.
Ear defenders.
Gloves, incorporating ballistic nylon lining on back of left hand.
Nylon leg guard.
Safety boots, incorporating inner ballistic nylon guard.

4 Allowances

Included in the times

a. For contingencies and work other than that described in paragraph 2 eg maintaining, refuelling and resharpening chainsaw, general preparation, unloading and setting up winch where not fixed to the Land Rover, fetching tools, moving to next site etc 30% of the cyclic time.

b. For personal needs and rest, 25% of the total working time.

To be included in the price per minute

c. The appropriate incentive allowance.

5 Method of selecting the time/output

Assess the mean tree volume (ie the volume of the average amount of usable timber) after the trees have been felled and ascertain the appropriate mean time or output from the table in paragraph 6. Usable timber in this context means that below 20–24 cm top diameter.

6 Output when felling diseased hedgerow elms

Mean tree volume (m³) (1)	Time per tree (minutes) (2)	Output per day for a team of 3 men (No of trees) (3)
0·2	81	17–18
0·3	98	14–15
0·4	112	12–13
0·5	125	11–12
0·6	138	10–11
0·7	148	9–10
0·8	156	9–10
0·9	163	8–9
1·0	169	8–9
1·1	174	
1·2	177	$7\frac{1}{2}$–$8\frac{1}{2}$
1·3	181	
1·4	184	7–8
1·5	187	7–8

7 Modifications and variations

N.B. These modifications and variations are only to be applied when the prevailing conditions or job specification differ from those listed in paragraphs 1 and 2.

They are best applied to the times in col 2 of table 6. To calculate output in number of trees, divide the resultant time into 1,440.

a. *Complete peeling*
Where complete peeling is necessary, add 10 minutes per m³.

b. *No peeling*
Where peeling is not included, reduce the times by 18%.

	Section	XVIII
	No	G8

Thinning and Clearfelling Broadleaves and Converting to Bar Lengths and Sudbrook Pulpwood
Forest of Dean

1 Conditions

The Output Guide applies to stands felled or thinned under the following conditions:

a. The average volume per tree determined by agreed methods.

b. Ground conditions average for locality; slopes up to 40% (22°); bracken and bramble up to 1 m high.

c. Racks at 20 m intervals in first thinnings.

d. One man working.

e. Take down normal for species and locality.

f. Extraction by forwarder or skidder.

2 Job specification

The Output Guide is for the following work:

a. Fell, chainsaw sned and crosscut the trees into 1·2 m or 2 m lengths as directed by the supervisor (diameter range 5 cm–30 cm). Butt lengths cut to a minimum length of 4 m, minimum top diameter 21 cm for hardwood bars.

b. Tops to be crosscut at 5 cm diameter, but on large trees, crown-wood to be converted when the shape and size fall within the product specification.

c. All cuts including the felling cut to be made squarely across the tree.

d. All buttresses to be trimmed by saw.

e. Ivy tendrils to be removed from produce.

f. Produce to be piled neatly at stump or in rack, with ends flush.

g. Stumps to be cut as low as possible.

h. Roads, rides, racks, main drains and produce piles to be cleared of lop and top. See paragraphs 7a and 7d.

i. In clearfelling areas, lop and top to be piled for burning or laid to assist forwarder extraction, as directed by the supervisor. See paragraphs 7b and 7c.

j. Unmeasurable trees to be felled and crosscut to ensure minimum interference with extraction. See paragraph 7e.

k. Count and check ('take-up') the number of pieces piled. See paragraph 7f.

3 Tools and equipment (including safety equipment)

a. A lightweight anti-vibration chainsaw complete with 38 cm cutter bar, tool kit and fuel cans.

b. Measuring stick marked for 1·2 m and 2 m lengths.

c. Alloy or plastic felling wedges.

d. Lightweight axe for trimming ivy.

e. Felt marker or timber crayon.

f. Notebook and pen or pencil.

g. Protective clothing viz:
Safety helmet BS 2826, 2095 or 5240.
Eye protection.
Ear defenders.
Gloves, incorporating ballistic nylon lining on back of left hand.
Nylon leg guard.
Safety boots, incorporating inner ballistic nylon guard.

4 Allowances

Included in the Output Guide

a. For contingencies and work other than that described in paragraph 2 eg tool maintenance, inspect and consider, fetching tools, walking to and from camp, clearing tops from drains and produce piles, on/off protective clothing, refuelling saw, supervisory visits, receiving pay, etc, etc—34% of the time spent felling, snedding, crosscutting and piling.

N.B. Repairs and maintenance of more than 15 minutes duration are not included.

b. For personal needs and rest—fell, sned, convert and stack 27% of cyclic time and 14% of other work. For personal needs and rest—fell, sned and off top 21% of cyclic time and 14% of other work.

5 Method of using the Output Guide

a. Entry to the tables in paragraph 6 is by the average volume per tree determined by agreed methods.

b. Where a known proportion of the trees in the stand are covered by ivy, the time per tree of given volume will have to be calculated on a proportional basis.

Example

In a stand of oak which is to be thinned, converted and stacked, the average volume per tree is 0·22 m³, 45% of the trees are ivy covered.

Mean tree volume m³	Standard time per tree without ivy	Standard time per tree with ivy
0·22	12·57	18·04

Calculation of time per tree

$$12\cdot57 \times \frac{55}{100} = 6\cdot91$$

$$18\cdot04 \times \frac{45}{100} = 8\cdot12$$

Time per tree 15·03 standard minutes

6 Time per tree in standard minutes

A Oak

Average volume (m³)	Trees without ivy		Trees with ivy	
	Fell, convert, stack	Fell only	Fell, convert, stack	Fell only
0·01	2·16	1·38	2·61	1·82
0·02	2·84	1·76	3·49	2·59
0·04	4·15	2·48	5·20	4·05
0·06	5·38	3·17	6·86	5·38
0·08	6·52	3·82	8·45	6·59
0·10	7·59	4·45	9·98	7·73
0·12	8·58	5·04	11·45	8·75
0·14	9·51	5·60	12·87	9·71
0·16	10·38	6·13	14·24	10·59
0·18	11·15	6·62	15·55	11·41
0·20	11·90	7·09	16·82	12·18
0·22	12·57	7·54	18·04	12·90
0·24	13·20	7·95	19·22	13·58
0·26	13·76	8·34	20·35	14·24
0·28	14·28	8·70	21·43	14·90
0·30	14·76	9·06	22·48	15·52
0·32	15·19	9·38	—	—
0·34	15·59	9·68	—	—
0·36	15·94	9·95	—	—
0·38	16·27	10·21	—	—
0·40	16·57	10·45	—	—
0·42	16·83	10·69	—	—

A Oak—continued

Average volume (m³)	Trees without ivy		Trees with ivy	
	Fell, convert, stack	Fell only	Fell, convert, stack	Fell only
0·44	17·08	10·90	—	—
0·46	17·31	11·09	—	—
0·48	17·53	11·26	—	—
0·50	17·73	11·44	—	—
0·52	17·91	11·60	—	—
0·54	18·09	11·74	—	—

N.B. Interpolate where necessary.

B Other broadleaves

Average volume m³	Beech understorey clearance	Beech thinning	Sweet chestnut clearfell
0·04	5·24	—	—
0·06	6·17	4·96	5·04
0·08	7·07	5·71	5·64
0·10	7·94	6·44	6·19
0·12	8·80	7·16	6·70
0·14	9·63	7·88	7·17
0·16	10·45	8·59	7·63
0·18	11·24	9·29	8·10
0·20	12·02	9·98	8·57
0·22	12·79	10·66	9·08
0·24	13·54	11·33	9·63
0·26	14·27	11·99	—
0·28	14·99	12·65	—
0·30	15·70	13·29	—
0·32	16·40	13·92	—
0·34	17·09	14·55	—
0·36	17·77	15·16	—
0·38	18·45	—	—
0·40	19·11	—	—
0·42	19·77	—	—
0·44	20·43	—	—

N.B. Interpolate where necessary.

7 Modifications and variations to the times

N.B. These modifications and variations are only to be applied when the prevailing conditions or job specification differ from those listed in paragraphs 1 and 2.

a. *Clearing lop and top from racks*
In first thinnings where it is necessary to clear lop and top from racks to facilitate extraction,
    Add 4% to the times.

b *Piling lop and top for burning*
In clearfelling areas where lop and top is piled ready for burning,
    Add up to 10% to the times.

c. *Laying lop and top in strips*
In areas where lop and top is laid in strips 10–20 m apart to assist the traction of the Volvo forwarder,
    Add 16% to the times.

d. *Removing lop and top from roads and rides*
When it is necessary to clear roads and rides of lop and top from edge trees,
    Add $2\frac{1}{2}$% to the times.

e. *Unmeasurable trees*
   (i) In first thinnings allow 0·50 standard minutes for each unmeasurable tree cut as per paragraph 2j.
   (ii) In second and subsequent thinnings and clearfelling allow 0·70 standard minutes for each unmeasurable tree as per paragraph 2j.

f. *'Taking up'*
When counting, checking and booking produce ('taking up') allow 2·3 standard minutes per m^3.

g. *Rough beech*
Where these are found in a matrix of first and second thinnings of oak, apply the standard minute values for beech understorey clearance given in paragraph 6b.

Section	XVIII
No	G10

Thinning, Crosscutting and Hand Piling for Skyline Winch or Forwarder Extraction Scots Pine for 3 m Pulpwood
North Scotland

1 Conditions

The Output Guide applies to stands thinned under the following conditions:

a. Moderately heavy selective thinning.

b. The brashing percentage at least 70% but see paragraph 7.

c. The ground conditions average for the species and locality ie slopes up to 18% (10°), normal amounts of brash, a few rocks and normal drainage patterns.

d. One-man working.

e. Average tree size less than 0.15m^3.

2 Job specification

The Output Guide is for the following work:

a. Fell, chainsaw sned and crosscut the trees into pulpwood billets which have the following specification: 0.05–0.41 m OB diameter range and a standard 3 m length with a downward tolerance to 2.4 m.

b. All stumps to be treated immediately after felling.

c. Billets to be piled (on a bearer for winch working) in the easiest direction for extraction. The piles to be no more than 0.5m^3 in size.

d. Racks to be kept clear of produce, as directed by the supervisor, for forwarder extraction.

e. Stumps to be cut as low as possible.

f. Branches to be cut off flush with the stem of the tree.

g. The maximum number of pulp billets to be obtained from each tree, within the specification limits given in 2a above.

h. Lop and top to be cleared from roads and main drains.

i. All cuts, including the felling cut to be made as squarely across the tree as possible.

j. All buttressing to be removed from the butt.

3 Tools and equipment (including safety equipment)

a. Lightweight chainsaw suitable for chainsaw snedding with spares, tool kit and fuel cans.

b. Plastic bottle for stump treatment.

c. Logger's tape or measuring rod.

d. Breaking bar or cant-hook.

e. Alloy wedge.

f. Protective clothing viz:
Safety helmet BS 2826, 2095 or 5240.
Eye protection.
Ear defenders.
Gloves, incorporating ballistic nylon lining on back of left hand.
Nylon leg guard.
Safety boots, incorporating inner ballistic nylon guard.

4 Allowances

Included in the Output Guide

a. For contingencies and work other than that performed on individual trees, eg refuelling, day-to-day maintenance on the saw, clearing brash from roads, stump treatment: 23% of the time spent thinning, snedding, crosscutting and stacking.

N.B. Mechanical breakdowns in excess of 15 minutes are not included in the allowances.

b. Personal needs and rest: 22% of the total working time.

To be included in the price per standard minute

c. The appropriate incentive allowance.

5. Method of selecting the standard minute value

a. Average billet volume is determined from a sample of 40 pieces selected at random over the thinning block.

b. The appropriate standard minute value or output is determined from the table in paragraph 6.

6 Standard minute values and outputs for thinning Scots pine in North Scotland—3 m pulpwood being the only product

Average volume per pulp billet (m³)	Standard minutes per pulp billet	Standard minutes per m³ of pulp	Output per working hour (m³)
0·020	2·46	123·00	0·488
0·021	2·46	117·14	0·512
0·022	2·46	111·82	0·537
0·023	2·46	106·96	0·561
0·024	2·46	102·50	0·585
0·025	2·46	98·40	0·610
0·026	2·54	97·69	0·614
0·027	2·54	94·07	0·638
0·028	2·54	90·71	0·661
0·029	2·54	87·59	0·685
0·030	2·54	84·67	0·709
0·031	2·62	84·52	0·710
0·032	2·62	81·88	0·733
0·033	2·62	79·39	0·756
0·034	2·62	77·06	0·779
0·035	2·62	74·86	0·801

7 Modifications and variations to the Output Guide

N.B. These modifications and variations are only to be applied when the prevailing conditions or job specification differ from those listed in paragraphs 1 and 2.

a. *Brashing percentage*

If the brashing percentage is between 25% and 70% add $2\frac{1}{2}$% to the standard minute values and reduce the output figure by a factor of 0·975. If the brashing percentage is between 0 and 25% add 7% to the standard minute values and reduce the output figure by a factor of 0·935.

Clearfelling Douglas Fir Shortwood Working
North Scotland

1 Conditions

This Output Guide applies to stands clearfelled under the following conditions:

a. The volume of the average tree is determined by the tariff system (see FC Booklet No 36).

b. The brashing percentage is at least 70%.

c. Forest floor conditions are average, ie slope no more than 40%, normal amounts of brash, a few rocks and normal drainage patterns.

d. One-man working.

2 Job specification

The Output Guide is for the following work:

a. Fell, chainsaw sned and crosscut the trees into logs (dimensions as directed by supervisor) and pulpwood of the following specification: 0·05–0·41 m OB diameter range and a standard 3 m length with a downward tolerance to 2·4 m. All logs produced are to be measured, the dimensions marked on one end and also recorded in a suitable notebook. If measuring, marking and booking is not required, see para 7a.

b. All stumps to be treated immediately after felling, see para 7b for the appropriate allowance.

c. Pulp billets to be piled (on a bearer for winch working) in the easiest direction for extraction. The piles to be no more than 0·5 m³ in size and free of brash.

d. Racks to be kept clear of produce, as directed by the supervisor, for forwarder extraction.

e. Stumps to be cut as low as possible.

f. Branches to be cut off flush with the stem of the tree.

g. The maximum volume of marketable produce to be obtained from each tree, within the specifications given in 2a above.

h. Lop and top to be cleared from roads and main drains.

i. All cuts, including the felling cut, to be made as squarely across the tree as possible.

j. All buttressing to be removed from the butt.

3 Tools and equipment (including safety equipment)

a. Lightweight chainsaw suitable for chainsaw snedding with spares, tool list and fuel cans.

b. Plastic bottle for stump treatment.

c. Logger's tape and diameter girth-tape.

d. Bilness breaking bar or cant-hook.

e. Alloy wedge.

f. Marking crayon, notebook and pencil.

g. Protective clothing viz:
Safety helmet BS 2826, 2095 or 5240.
Eye protection.
Ear defenders.
Gloves, incorporating ballistic nylon lining on back of left hand.
Nylon leg guard.
Safety boots, incorporating inner ballistic nylon guard.

4 Allowances

a. For contingencies and work other than that performed on individual trees, eg refuelling, day-to-day maintenance on the saw, clearing brash from roads: 18% of the time spent felling, snedding, crosscutting, stacking, measuring, marking and booking.

N.B. Mechanical breakdowns in excess of 15 minutes are not included in the allowances.

b. Personal needs and rest: 22% of the total working time.

5 Method of using the Output Guide

a. The trees shall be tariffed according to FC Booklet No 36.

b. The average tree volume will be calculated and used as the point of entry to table 6.

6 Time in standard minutes for clearfelling DF—shortwood working

Average tree m^3	Standard minutes per tree	Standard minutes per m^3	Output of logs and pulp m^3 per working hour
0·7	14·68	20·97	2·86
0·8	15·74	19·68	3·05
0·9	16·77	18·63	3·22
1·0	17·78	17·78	3·37
1·1	18·75	17·05	3·52
1·2	19·70	16·42	3·65
1·3	20·63	15·87	3·78
1·4	21·52	15·37	3·90
1·5	22·39	14·93	4·02
1·6	23·23	14·52	4·13
1·7	24·04	14·14	4·24
1·8	24·83	13·79	4·35
1·9	25·59	13·47	4·46
2·0	26·32	13·16	4·56

7 Modifications and variations to the Output Guide

N.B. These modifications and variations are only to be applied when the prevailing conditions or job specification differ from those listed in paragraphs 1 and 2.

a. *Measuring, marking and booking of sawlogs*
   If this work is not required, deduct 15% from the times in paragraph 6 and increase output figures accordingly.

b. *Stump treatment*
   For preparation of the solution, filling the application bottle, applying the liquid to cover completely all cut stumps, add 3% to the times in paragraph 6 and reduce output figures accordingly.

1 Conditions

The Output Guide applies to windblown trees cleared under the following conditions:

a. Trees blown singly, in groups or snapped.

b. Tree volume determined by a locally agreed method.

c. Forest floor conditions are average, ie generally flat or with a slope of not more than 10% (6°) and with some decayed brash and light undergrowth.

d. The brashing percentage is at least 90%.

e. Supplies of fuel, oil and stump treatment solution readily available at convenient places.

f. One-man working but a tractor, complete with driver and fitted with tongs, to be available to each 2 or 3 fellers for pulling tangled trees apart.

2 Job specification

The Output Guide is for the following work:

a. Stumps to be cut as low as safely possible.

b. Branches to be cut off flush with the stem and tops to be cut off at a diameter specified by the supervisor.

c. Primary conversion into sawlog(s) and remaining pole to be carried out at the time of clearing.

d. All cuts to be made square across the stem.

e. Butts to be squared off where necessary.

f. Rides and roads to be kept clear of lop and top.

g. Stumps to be treated as soon as possible after cutting. Care should be taken not to taint sawlogs when the root does not fall away from the butt. (Extra time is required for stump treatment—see paragraph 7.)

3 Tools and equipment (including safety equipment)

a. Lightweight AV chainsaw, appropriate maintenance tools and fuel cans.

b. Breaking bar where required.

c. Logger's belt and spring loaded tape.

d. Stump treatment equipment.

e. Protective clothing viz:
Safety helmet BS 2826, 2095 or 5240.
Eye protection.
Ear defenders.
Gloves, incorporating ballistic nylon lining on back of left hand.
Nylon leg guard.
Safety boots, incorporating inner ballistic nylon guard.

4 Allowances

The following allowances are included in the Output Guide:

a. For contingencies and work other than that actually performed on individual trees, eg refuelling saws, to and from work site, clearing tops from rides, supervisory visits, etc, 22% of the time spent on 'felling' and snedding. This has been increased to 40% to take account of the extra saw sharpening entailed in windblown areas and for unavoidable delays caused by having tangled trees pulled apart by the tractor.

b. For personal needs and rest, 22% of the total working time.

5 Method of using the Output Guide

The average tree volume, number of whorls to be snedded per tree and the number of pieces to be cut per tree should be determined by an agreed method. The time per tree is then taken from the tables in paragraph 6 according to species.

Example 1
CP.
$0.50\,m^3$ per tree.
22 whorls per tree.
2·50 pieces cut per tree.

Fell and sned   = 8·92 SMs per tree (from table 6B col II).

Treat stump   = 8·92 SMs × 5% (from paragraph 7)
      = 0·45 SMs per tree.

Measure and crosscut = 1·01 SMs per tree (from table 6B col V).

Total time per tree   = 8·92 + 0·45 + 1·01 SMs
      = 10·38 SMs.

Example 2

CP.

0·50 m³ per tree.

23 whorls per tree.

2·30 pieces cut per tree.

Fell and sned $\quad = 8·92 + 0·14 - (0·18 \times 2)$ SMs per tree (from table 6B col II with addition for extra whorl and deduction for fewer pieces cut)

$\quad\quad\quad\quad\quad\quad\quad = 8·70$ SMs per tree.

Treat stump $\quad = 8·92 \times 5\%$ (from paragraph 7)

$\quad\quad\quad\quad\quad\quad\quad = 0·45$ SMs per tree.

Measure and crosscut $= 1·01 - (0·06 \times 2)$ SMs per tree (from table 6B col V with deduction for fewer pieces cut)

$\quad\quad\quad\quad\quad\quad\quad = 0·89$ SMs per tree.

Total time per tree $\quad = 8·70 + 0·45 + 0·89$ SMs

$\quad\quad\quad\quad\quad\quad\quad = 10·04$ SMs.

6A   Scots pine—time in standard minutes per tree

Volume in m^3 per tree	Time to fell and sned in SMs per tree	Mean number of pieces cut per tree	Mean number of whorls per tree	Time to measure and crosscut in SMs per tree
I	II	III	IV	V
0·20	5·09	2·1	18	0·51
0·25	5·50	2·2	18	0·63
0·30	5·98	2·3	19	0·73
0·35	6·54	2·4	19	0·82
0·40	7·10	2·5	20	0·90
0·45	7·69	2·6	20	0·97
0·50	8·25	2·7	20	1·04

(i) *Add/subtract* 0·14 SMs per whorl to the time shown in col II if the mean number of whorls per tree is different to that shown in col IV.

(ii) *Add/subtract* 0·06 SMs for each tenth piece to the time shown in col II if the mean number of pieces per tree is different to that shown in col III.

(iii) *Add/subtract* 0·02 SMs for each tenth piece to the time shown in col V if the mean number of pieces is different to that shown in col III.

6B Corsican pine—time in standard minutes per tree

Volume in m³ per tree	Time to fell and sned in SMs per tree	Mean number of pieces cut per tree	Mean number of whorls per tree	Time to measure and crosscut in SMs per tree
I	II	III	IV	V
0·25	6·59	2·2	20	0·67
0·30	7·21	2·3	20	0·73
0·35	7·75	2·3	21	0·80
0·40	8·20	2·4	21	0·87
0·45	8·59	2·5	22	0·94
0·50	8·92	2·5	22	1·01
0·55	9·17	2·5	23	1·06
0·60	9·38	2·6	23	1·11
0·65	9·55	2·6	23	1·16
0·70	9·67	2·6	24	1·21
0·75	9·75	2·6	24	1·25
0·80	9·80	2·6	24	1·30
0·85	9·85	2·6	24	1·33
0·90	9·87	2·5	24	1·37
0·95	9·89	2·5	24	1·38
1·00	9·91	2·5	25	1·42

(i) *Add/subtract* 0·14 SMs per whorl to the time shown in col II if the mean number of whorls per tree is different to that shown in col IV.

(ii) *Add/subtract* 0·18 SMs for each tenth piece to the time shown in col II if the mean number of pieces per tree is different to that shown in col III.

(iii) *Add/subtract* 0·06 SMs for each tenth piece to the time shown in col V if the mean number of pieces is different to that shown in col III.

7 Modifications and variations to the Output Guide

N.B. These modifications and variations are only to be applied when the prevailing conditions or job specification differ from those listed in paragraphs 1 and 2.

a. *Stump treatment*

For treatment of stumps at the time of 'felling'

Add 5% to the time shown in col II.

N.B. The method of making this addition is shown in the examples in paragraph 5.

Clearfelling Norway and Sitka Spruce ('Scoot' Felling)
North East England

1 Conditions

The Output Guide applies to stands clearfelled under the following conditions:

a. The volume of the average tree determined by the tariff system. In stands previously crown thinned it may be necessary to sub-divide the DBH distribution (see FC Booklet No 36, paragraph 26 (ix) page 12); in these cases a different average tree size should be used for each subdivision when using the Output Guide.

b. The brashing percentage is at least 90% of the measurable trees (see paragraph 7b).

c. Floor conditions are average for the species, ie normal amounts of lop and top from thinnings, little ground vegetation, few rocks, normal drainage patterns (see paragraph 7a).

d. One-man working.

e. Slopes of up to 20% (11°).

2 Job specification

The Output Guide applies to the following work:

a. Pole length working.

b. Trees to be felled in 'scoots' or drifts in the direction most suitable for extraction, 'scoot' width to be an average of five trees.

c. Stumps to be cut as low as possible.

d. Branches, brashing and pruning knots to be cut off flush with the stem of the tree, snedding with the chainsaw.

e. Lop and top to be cleared from major racks, roads and main drains as required by the supervisor.

f. Tops to be cut off at a diameter of 7·5 cm (for greater or lesser cut off diameters see paragraph 7e).

g. Tops to be cut into lengths not exceeding 1·2 m.

h. All cuts, including the felling cut, to be made as squarely across the tree as possible.

i. Buttresses to be removed with the chainsaw after felling.

j. Stumps to be treated immediately after felling.

N.B. Treating stumps requires extra time—see paragraph 7d.

3 **Tools and equipment** (including safety equipment)

a. Lightweight anti-vibration chainsaw of an approved pattern and suitable for chainsaw snedding, together with spares and maintenance tools.

b. Breaking bar which may have a cant-hook attachment.

c. 13 cm plastic felling wedges as required.

d. When larger trees are turned from the butt, or timber pieces are turned after crosscutting, a cant-hook or breaking bar with cant-hook attachment (see 3b above).

N.B. Time for crosscutting is not included in the Output Guide.

e. If trees are crosscut at stump, Spencer or Bushman's logger's tape and pulphook.

N.B. Time for crosscutting is not included in the Output Guide.

f. Suitable equipment for stump treatment.

g. Protective clothing viz:
Safety helmet BS 2826, 2095 or 5240.
Eye protection.
Ear defenders.
Gloves, incorporating ballistic nylon lining on back of left hand.
Nylon leg guard.
Safety boots, incorporating inner ballistic nylon guard.

4 **Allowances**

a. *Included in the times*

(i) For contingencies and work other than that performed on individual trees, eg refuelling and day to day maintenance of the saw, clearing lop and top from rides etc 23% of the time spent on clearfelling.

N.B. Mechanical breakdowns to the saw in excess of 15 minutes are not included in the allowance and should be paid for at time rates.

(ii) For personal needs and rest, 25% of the total working time.

b. For calculating the payment for worker-owned chainsaws, chainsaw utilisation is 70% of the total working time when the chainsaw is running during stump treatment or when stumps are not treated.

When the saw is not running during stump treatment chainsaw utilisation is 70% of the standard time for felling or about 65% of the total working time.

5 Method of using the Output Guide

a. Trees should be tariffed by the method described in FC Booklet No 36.

b. To select the time from the tables use the appropriate average volume for the stand and read off the time per tree.

Example 1

Norway spruce.

Average tree volume $= 0 \cdot 16 \, m^3$.

Conditions and job specification as shown in paragraphs 1 and 2.

Fell and sned (from paragraph 6a) = 6·88 standard minutes per tree.

Treat stump (from paragraph 7d) = 0·40 standard minutes per tree.

Total = 7·28 standard minutes per tree.

Example 2

Norway spruce.

Average tree volume $= 0 \cdot 27 \, m^3$.

Conditions and job specification as shown in paragraphs 1 and 2.

Fell and sned (from paragraph 6a).

Interpolated: $8 \cdot 78 + \left[\dfrac{(9 \cdot 59 - 8 \cdot 78)}{4} \times 3 \right]$

$= 9 \cdot 39$ standard minutes per tree.

Treat stump (from paragraph 7d) = 0·60 standard minutes per tree.

Total = 9·99 standard minutes per tree.

6 Output Guide for 'scoot' felling

A *Norway spruce*

Volume of average tree in m^3	Time per tree in standard minutes
0·04	3·05
0·06	3·79
0·08	4·48
0·10	5·14
0·12	5·75
0·14	6·33
0·16	6·88
0·18	7·39
0·20	7·88
0·24	8·78
0·28	9·59
0·32	10·33
0·36	11·03
0·40	11·70
0·44	12·37

N.B. Interpolate as necessary.

B *Sitka spruce*

Volume of average tree in m^3	Time per tree in standard minutes
0·04	3·50
0·06	4·12
0·08	4·71
0·10	5·29
0·12	5·84
0·14	6·37
0·16	6·88
0·18	7·37
0·20	7·79
0·24	8·71
0·28	9·56
0·32	10·32
0·36	11·03
0·40	11·68
0·44	12·28
0·48	12·84
0·52	13·35
0·56	13·84
0·60	14·30
0·64	14·72
0·68	15·12
0·72	15·51
0·76	15·88

N.B. Interpolate as necessary.

7 Modifications and Variations to the Output Guide

N.B. These modifications and variations are only to be applied when the prevailing conditions, or job specification, differ from those listed in paragraphs 1 and 2.

a. *Ground conditions*

Where ground conditions are difficult because of rocks, deep drains or excessive steepness (eg slopes in excess of 20% (11°))
Add up to 10% to the times in steps of $2\frac{1}{2}$%.

b. *Brashing percentage*

If the percentage of trees brashed is less than 90% the brashing of all measurable unbrashed trees should be paid for. The time for chainsaw brashing (or minimal brashing and extra snedding) is 0·75 standard minutes per tree.

c. *Unmeasurable trees*

If unmeasurable trees are felled and cut up into 1·2 m lengths the time for this operation is 1·00 standard minute per stem.

N.B. Extra brashing and the disposal of unmeasurable trees may be paid for separately from the felling of the crop. Alternatively, the total time for brashing and/or the felling and disposal of unmeasurable trees may be worked out and divided by the number of the crop trees, the resulting figure being added to the time per tree.

d. *Stump treatment*

For applying nitrite, polybor or urea to the stump immediately after felling (including fetching chemicals, filling cans, etc)

Add 0·40 standard minutes per tree for trees up to 0·18 m³

0·60 standard minutes per tree for trees over 0·18 m³ to 0·72 m³

0·80 standard minutes per tree for trees over 0·72 m³.

e. *Trees snedded to top diameter other than 7·5 cm*

When cutters are required to sned poles to top diameters other than 7·5 cm modifications should be made to the times as shown in the table below:

Snedding from 7·5 cm top diameter to	*Time per tree in standard minutes*
9 cm top diameter	Subtract 0·46 SMs per tree
8 cm top diameter	Subtract 0·15 SMs per tree
7 cm top diameter	Add 0·15 SMs per tree
6 cm top diameter	Add 0·46 SMs per tree

N.B. These times should not be used for trees averaging more than 0·20 m³ in volume.

Thinning, Crosscutting and Hand Piling for Skyline Winch or Forwarder Extraction of Sitka and Norway Spruces for Scottish Pulp and Logs
North Scotland

1 Conditions

The Output Guide applies to stands thinned under the following conditions:

a. Moderately heavy selective thinning or, in the case of first thinnings only, a chevron or herringbone pattern of marking.

b. The brashing percentage at least 70% but see paragraph 7a.

c. The ground conditions average for the species and locality ie slope up to 20° (37%), normal amounts of brash, a few rocks and normal drainage patterns. In more difficult conditions see paragraph 7b.

c. One-man working.

2 Job specification

The Output Guide is for the following work:

a. Fell, chainsaw sned and crosscut the trees into logs (dimensions as directed by supervisor up to a maximum length of 6m) and Scottish pulpwood which has the following specification: 0·05–0·41 m OB diameter range and a standard 3 m length with a downward tolerance of 2·4m. If the logs produced are to be measured, marked and the details booked see paragraph 7c. If light pokers (length 3·6–4·3 m maximum butt diameter 0·06 m) are required see paragraph 7d.

b. All stumps to be treated with urea immediately after felling (see paragraph 7e).

c. Pulp billets to be piled (on a bearer for winch working) in the easiest direction for extraction. The piles to be no more than 0·5 m³ in size and free of brash.

d. Racks to be kept clear of produce, as directed by the supervisor, for forwarder extraction.

e. Stumps to be cut as low as possible.

f. Branches to be cut off flush with the stem of the tree.

g. The maximum volume of marketable produce to be obtained from each tree, within the specifications given in 2a above.

h. Lop and top to be cleared from roads and main drains.

i. All cuts, including the felling cut, to be made as squarely across the tree as possible.

j. All buttressing to be removed from the butt.

3 Tools and equipment (including safety equipment)

a. Lightweight chainsaw suitable for chainsaw snedding with spares, tool kit and fuel cans.

b. Plastic bottle for stump treatment.

c. Logger's tape or measuring rod.

d. Breaking bar or cant-hook.

e. Alloy wedge.

f. Protective clothing viz:
Safety helmet BS 2826, 2095 or 5240.
Eye protection.
Ear defenders.
Gloves incorporating ballistic nylon lining on back of left hand.
Nylon leg guard.
Safety boots incorporating inner ballistic nylon guard.

4 Allowances

Included in the Output Guide

a. For contingencies and work other than that performed on individual trees, eg refuelling, day-to-day maintenance on the saw, clearing brash from roads: 22% of the time spent thinning, snedding, crosscutting and stacking.

N.B. Mechanical breakdowns in excess of 15 minutes are not included in the allowances.

b. Personal needs and rest: 26% of the total working time.

To be included in the price per standard minute

c. The appropriate incentive allowance.

5 Method of selecting the standard minute value

a. Average pulp volume is determined from a sample, containing a fixed number of pulp pieces, selected at random over the thinning area. Sample size is 36 pieces in all cases, no matter whether pulp pieces only or both pulp pieces and sawlogs are produced. Similarly average sawlog volume is determined from a sample of 16 pieces selected at random over the thinning area. In some circumstances less than this number of logs will be produced from any one thinning area. In such cases all sawlogs should be measured and the true average volume determined.

b. The appropriate standard minute value or output is determined from the tables in paragraph 6.

6A   Standard minute values and outputs for thinning Sitka spruce

(i)   *Pulpwood*

Average volume per pulp billet m^3	Standard minutes per pulp billet	Standard minutes per m^3 of pulp produced	Output of pulp in m^3 per working hour
0·020	3·24	162·00	0·370
0·025	3·37	134·80	0·445
0·030	3·46	115·33	0·520
0·035	3·56	101·71	0·590
0·040	3·62	90·50	0·663
0·045	3·74	83·11	0·722
0·050	3·80	76·00	0·789
0·055	3·85	70·00	0·857
0·060	3·88	64·67	0·928
0·065	3·88	59·69	1·005

(ii)   *Sawlogs*

Average volume per sawlog m^3	Standard minutes per sawlog	Standard minutes per m^3 of sawlogs produced
0·25	4·53	18·12
0·27	5·48	20·30
0·29	6·43	22·17
0·31	7·36	23·74
0·33	8·31	25·18
0·35	9·25	26·43

6B Standard minute values and outputs for thinning Norway spruce

(i) *Pulpwood*

Average volume per pulp billet m^3	*Standard minutes per pulp billet*	*Standard minutes per m^3 of pulp produced*	*Output of pulp in m^3 per working hour*
0·020	2·67	133·50	0·449
0·025	2·92	116·80	0·514
0·030	3·17	105·67	0·568
0·035	3·42	97·71	0·614
0·040	3·62	90·50	0·663
0·045	3·74	83·11	0·722
0·050	3·80	76·00	0·789
0·055	3·85	70·00	0·857
0·060	3·88	64·67	0·928
0·065	3·88	59·69	1·005

(ii) *Sawlogs*

Insufficient data is available for the publication of a Norway spruce sawlog table. The limited information available indicates that outputs are very similar to those given for Sitka spruce.

7 Modifications and variations to the Output Guide

N.B. These modifications and variations are only to be applied when the prevailing conditions or job specification differ from those listed in paragraphs 1 and 2.

a. *Brashing percentage*

If the brashing percentage is between 25% and 70% add 5% to the standard minute values. If the brashing percentage is between 0 and 25% add 7·5% to the standard minute values.

b. *Ground conditions*

Where the slope is between 20° (37%) and 45° (100%) add up to 25%, in steps of 5%, to the standard minute values.

c. *Measuring, marking and booking of sawlogs*

For measuring a log, marking the results on the log, and booking the figures allow 1·72 standard minutes per log.

d. *Light pokers*

For the production of light pokers from unmeasurable trees and tops allow 2·73 standard minutes per poker cut and stacked.

e. *Stump treatment*

For the preparation of the urea solution, filling the application bottle, applying the liquid to cover completely all cut stumps add 4·5% to the standard minute values.

N.B. Output figures should be reduced in proportion to the increase of standard minute values.

Row and Chevron* Thinning Sitka Spruce
Wales

Note:

a. Chevron thinning as used here includes staggered chevron thinning.

b. For selective thinning Standard Time Table XVIII/22 should continue to be used.

1 Conditions

The Output Guide applies to crops thinned under the following conditions:

a. First row or chevron thinning. In brashed stands the thinning may follow a selective thinning.

b. The crops may be unbrashed (table 6A) or brashed (table 6B) where 85% or more of measurable trees are brashed. In partially brashed and incompletely brashed crops table 6B applies, with modifications from paragraph 7c.

c. Ground conditions normal for the species in Wales, viz: up to 30% (17°) slope, few rocks and normal drainage pattern. For more difficult conditions see paragraph 7d.

d. Row thinning and chevron main racks to be one row of trees wide, or an equivalent distance across the rows. (For double row or double row main rack chevron thinning see paragraph 7a.)

e. Trees readily identified (eg in rows, or marked) before thinning. If the operator has to align side racks during thinning this work should be allowed for separately, see paragraph 7e.

f. Take down problems normal. For difficult take down see paragraph 7f.

g. One-man working using a chainsaw for felling and snedding.

h. Working method as described in the Appendix.

i. Mean tree volume to be determined by the tariff system (see FC Booklet No 36) or by any other recognised method.

2 Job specification
The Output Guide is for the following work:

a. Stumps to be cut as low as possible.

b. Cuts, including the felling cut, to be made squarely across the pole.

c. Branches and pruning knots to be cut off flush with the stem of the trees.

d. All buttressing to be removed.

e. Tops to be cut off at a diameter of about 7 cm. If snedding is required to a smaller top diameter the extra work should be allowed for, see paragraph 7g.

f. Tops and some branches to be disposed of to the side of or out of the rack. If almost complete removal of branches is necessary the extra work should be allowed for, see paragraph 7i.

g. Trees to be felled for tip first extraction unless otherwise stipulated by the supervisor. For uphill felling see paragraph 7f.

h. Lop and top to be cleared from roads and main drains.

i. Stumps to be treated immediately after felling. This work should be allowed for separately, see paragraph 7j.

j. Where stacking of smaller poles ready for chokering is required the work should be allowed for separately, see paragraph 7k.

3 Tools and equipment (including safety equipment)

a. A lightweight anti-vibration chainsaw of an approved type, suitable for chainsaw snedding, and tool kit for chainsaw maintenance.

b. Fuel and oil cans.

c. Breaking bar with cant-hook attachment.

d. Alloy or plastic felling wedges.

e. Stump treatment equipment.

f. Protective clothing viz:
Safety helmet BS 2826, 2095 or 5240.
Eye protection.
Ear defenders.
Gloves, incorporating ballistic nylon lining on back of left hand.
Nylon leg guard.
Safety boots, incorporating inner ballistic nylon guard.

4 Allowances

a. The following allowances are included in the Output Guide:
(i) For contingencies and work other than that performed on individual trees, eg refuelling and day-to-day maintenance of the saw, clearing lop and top from roads and drains,

fetching tools and supplies, etc, 28% of the time actually spent felling.

N.B. Mechanical breakdowns to the saw in excess of 15 minutes are not included in the allowance and should be paid for separately.

(ii) For personal needs and rest 25% of the total working time.

b. Payment for worker owned chainsaws should be based on chainsaw utilisation of 70% of the total time.

5 Method of using the Output Guide

a. Select the table appropriate to the thinning method, as follows:
Unbrashed crops, row thinning table 6A col (ii).
Unbrashed crops, chevron thinning, table 6A col (iii).
Brashed crops, row and chevron thinning, table 6B.

b. Obtain mean tree volume by tariffing or by any other recognised method.

c. Read the time per tree from the table, and modify as necessary with reference to paragraph 7.

Example 1
Single row thinning
Crop 70% brashed
Mean tree volume 0·08 m³
Other conditions and job specification as shown in paragraphs 1 and 2

Time per tree (from table 6B)	= 6·3 SMs per tree
Addition for chainsaw brashing (from paragraph 7c)	= 0·14 SMs per tree
Allowance for stump treatment (from paragraph 7j)	= 0·30 SMs per tree
Total time	= 6·74 SMs per tree

Example 2
Chevron thinning
Unbrashed crop
Single row main rack
Uphill felling
50% of trees turned at the butt
Mean pole volume 0·125 m³
Other conditions and job specification as shown in paragraphs 1 and 2
Time per tree (from table 6A col (iii))

Interpolated: $11·3 + \left[\dfrac{(11·5 - 11·3)}{2}\right]$ = 11·4 SMs per tree

Allowance for uphill felling (from paragraph 7f) = 0·2 SMs per tree
Allowance for turning at butt (from paragraph 7h)

Interpolated: $0·55 + \left[\dfrac{(0·65 - 0·55)}{2}\right]$ = 0·6 SMs per tree

But only 50% of trees require turning at the butt
∴ allowance = 0·6 × 50%   =   0·3 SMs per tree
Allowance for stump treatment (from
paragraph 7j)   =   0·4 SMs per tree
Total time = 11·4 + 0·2 +
0·3 + 0·4
= 12·3 SMs per tree

Example 3
Double row thinning
Unbrashed crop
Mean tree volume 0·09 m³
Difficult terrain
Other conditions and job specification as shown in paragraphs 1 and 2
Time per tree (from paragraph 7a)   = 8·4   SMs per tree
5% allowance for difficult terrain (from
paragraph 7d)   = 0·42 SMs per tree
Allowance for stump treatment   = 0·30 SMs per tree
Total time = 9·12 SMs per tree

6A   Time in standard minutes per tree for thinning unbrashed crops

Average tree volume m³ (i)	Row thinning (ii)	Chevron thinning (iii)
0·04	5·5	7·2
0·05	6·0	8·0
0·06	6·5	8·7
0·07	7·0	9·3
0·08	7·4	9·8
0·09	7·8	10·2
0·10	8·2	10·6
0·11	8·5	11·0
0·12	8·8	11·3
0·13	9·1	11·5
0·14	9·3	11·7
0·15	9·5	11·8
0·16	9·7	12·0
0·17	9·9	12·1
0·18	10·0	12·3

Interpolate as necessary.

For double row or double row main rack chevron
thinning in unbrashed crops, see paragraph 7a.

6B Time in standard minutes per tree for thinning brashed crops

N.B. This table refers to single row or chevron with single row main racks.

Average tree volume m³	Time per tree in standard minutes
0·04	4·9
0·05	5·3
0·06	5·6
0·07	6·0
0·08	6·3
0·09	6·7
0·10	7·0
0·11	7·3
0·12	7·6
0·13	7·9
0·14	8·2
0·15	8·5
0·16	8·7
0·17	9·0
0·18	9·3
0·19	9·5
0·20	9·7
0·21	10·0
0·22	10·2
0·23	10·4
0·24	10·6

Interpolate as necessary.

No information is available for double row or double row main rack chevron thinning in brashed crops.

7 Modifications and variations to the Output Guide

N.B. These modifications and variations are only to be applied when the prevailing conditions or job specification differ from those listed in paragraphs 1 and 2.

a. *Double row or double row main rack chevron thinning in unbrashed crops*

Where double row or double row main rack thinning is carried out in **unbrashed** crops use the table on page 236.

Tree volume (m^3)	Double row	Chevron double row main rack
0·04	5·9	5·8
0·05	6·5	6·5
0·06	7·1	7·2
0·07	7·6	7·8
0·08	8·0	8·2
0·09	8·4	8·7
0·10	8·8	9·1
0·11		9·5
0·12		9·9
0·13		10·2
0·14	No information available	10·5
0·15		10·9
0·16		11·0
0·17		11·3
0·18		11·4

b. *Unmeasurable trees*

The time taken to dispose of unmeasurable trees is not included in the times in table 6A or 6B.

When unmeasurable trees are felled, cut up if necessary, and disposed of at the side of the rack or in adjoining rows, the time for this operation is:
(i)   In unbrashed stands 0·9 SMs per tree.
(ii) In brashed stands   0·8 SMs per tree.

When unmeasurable trees are felled and snedded to 3 or 4cm top diameter, the time for this operation is:
(i)   In unbrashed stands 4·0 SMs per tree.
(ii) In brashed stands   3·6 SMs per tree.

N.B. Unmeasurable trees usually need brashing whether in brashed or in unbrashed stands. The time for snedded trees includes brashing time but not stacking.

c. *Chainsaw brashing in brashed stands*

Time for brashing is included with the felling times in table 6A, when table 6B is used an addition for chainsaw brashing is necessary. The extra time shown below should be added to every tree—see paragraph 5c Example 1.

N.B. **These additions are for use with table 6B only.**

Brashing percentage	Addition to each tree
80	0·05 SMs
70	0·14 SMs
60	0·23 SMs
50	0·32 SMs

N.B. In crops where the brashing intensity is less than 50% it is advisable to work out separate times for brashed and unbrashed crops and to calculate a weighted average in proportion to the brashing intensity prevailing.

d. *Difficult terrain*
   Where ground conditions are difficult because of excessive steepness, deep drains, exceptionally rocky ground, etc
   Add up to 10% (in 5% steps).

e. *Positioning side racks in chevron thinning*
   If side racks have not been marked out in advance and this is done by the operator during thinning
   Add one standard minute per side rack.

f. *Difficult take down*
   (i)   In single row thinning, or in side racks of similar width, in badly delayed thinning or in very closely planted or uneven crops
         Add up to 5% of the felling time.
   (ii) For uphill felling
         Add 0·2 standard minutes per tree.

g. *Cutting to top diameter of less than 7 cm*
   If the operator is *requested* to sned trees to top diameters smaller than 7 cm the extra snedding should be allowed for as follows:

Top diameter	Addition per tree
3	0·90 SMs
4	0·70 SMs
5	0·50 SMs
6	0·30 SMs

h. *Turning trees at butt*
   In most areas it is normal for *some* of the larger trees to be turned at the butt using a breaking bar/cant-hook and this is allowed for in the times. On difficult areas, however, it is sometimes necessary to turn up to *half* of the trees at the butt and in such

areas the trees turned at butt should be allowed for as follows, interpolating as necessary:

Mean tree volume	Addition for each tree turned
0·05 m³	0·45 SMs
0·10 m³	0·55 SMs
0·15 m³	0·65 SMs
0·20 m³	0·75 SMs
0·25 m³	0·85 SMs

i. *Clearing branches from racks*

If the operator is *requested* to move almost all of the branches out of the rack during felling, this extra work should be allowed for

Add 0·10 standard minutes per tree.

j. *Stump treatment*

The following *additions* should be made for applying a fungicide to the stumps immediately after felling:

Mean tree volume	Time per tree for stump treatment
Less than 0·10 m³	0·30 SMs
0·10 m³–0·149 m³	0·40 SMs
0·15 m³–0·199 m³	0·50 SMs
0·20 m³ and over	0·60 SMs

k. *Stacking poles*

Stacking the snedded poles into twos, threes and fours with their tips together for chokering is standard practice in many forests but has not been included in the Output Guide.

It is usual to move one or more small poles to a larger pole which is not moved and the number of poles moved to stacks for chokering may be obtained in each rack by deducting the number of stacks from the number of poles.

For each pole moved

Add 0·25 standard minutes.

Working method

The normal working method to which the Output Guide applies is as follows:

(i)   Walk to tree.

(ii) Chainsaw brash unbrashed or partially brashed trees to shoulder height (see Note (a)).

(iii) Lay in (if necessary).

(iv) Saw through.

(v)   Take down.

(vi) Treat stump.

(vii) Trim butt (when necessary).

(viii) Sned, working up the tree, cutting off all the branches which can be reached from one side of the tree.

(ix) Cut off top (and cut up top when necessary).

(x)   Dispose of top at side of rack or in adjacent row.

(xi) Turn tree and position tip ready for extraction (trees small enough for stacking—turn tree only) (see Note (b)).

(xii) Finish snedding, working back down the tree.

(xiii) Stack pole, if small enough, ready for chokering (see Note (c)).

Notes:

(a) It is possible to fell after brashing just enough branches to allow access to the tree. The method described above does not increase the total time for the tree.

(b) Trees too big for turning at the tip by hand may sometimes be turned using a branch left for the purpose. If many trees need to be turned at the butt using a breaking bar/cant-hook, extra time is required.

(c) Stacking for chokering is usually done by moving smaller poles to bigger ones. Tips of bigger trees may sometimes be brought level ready for chokering by varying the cut off diameter.

(d) In chevron thinning it is advisable, particularly for pole length ground skidding extraction, to fell and extract the main rack trees before felling the side rack trees. This method makes a marginal difference in the average time per pole (fractions of a minute saved) and is recommended because it makes working conditions much easier and probably safer.

Clearing Windblown Sitka Spruce, Norway Spruce and Douglas Fir
Pole Length Working (and Partial Conversion)
Wales and Marches

1 Conditions

The Output Guide applies to windblown trees cleared under the following conditions:

a. Trees are blown singly, in small or large groups, or snapped.

b. Tree volumes are determined by a recognised and locally agreed method.

c. Brashing percentage is at least 90%, but see paragraph 7b.

d. Forest floor conditions are average for the species, ie normal amounts of lop and top from thinnings, little ground vegetation, few rocks, normal drainage patterns, slopes up to 20% (11°), but see paragraph 7a.

e. Trees are snedded to *safe* limits in situ. Any tree not snedded, or not completely snedded in situ, is primary extracted on to cleared ground where snedding can be completed.

f. One-man working using the methods described in the Appendix.

N.B. It is assumed that an extraction machine and an operator(s) is available for the primary extraction of trees which are difficult or unsafe to work in situ. Approximate gang balance—one extraction machine per two or three fellers.

2 Job specification

The Output Guide is for the following work:

a. Pole length working where possible in safety. For marking and measuring sawlogs see paragraph 7c. Large trees may require primary conversion at stump. For marking, measuring and cross-cutting sawlogs during snedding see paragraph 7d. (Each sawlog is completely snedded.)

b. Stumps to be cut as low as possible in safety and root plates left in a safe condition (ie ensuring they will not fall back at an inconvenient moment).

c. Branches to be cut off flush with the stem. Smaller trees will be turned from the tip for snedding the undersides. Larger trees (above 0·25 m³) will normally only be snedded as thoroughly as is possible without turning them.

d. Tops to be cut off at about 7 cm diameter and cut up into lengths not exceeding 1·2 m as required by the supervisor.

e. Trees standing within or bordering on the windblown area to be felled as directed by the supervisor. (If the number of standing trees exceeds 20% of the windblown trees, the standard time tables for clearfelling should be used to obtain times for those trees).

f. All cuts to be made squarely across the stem and butts to be squared off where necessary.

g. Roads and rides to be kept clear of lop and top, as directed by the supervisor.

h. Stumps to be treated as soon as possible after cutting. (Extra time is required for stump treatment—see paragraph 7e.)

3 Tools and equipment (including safety equipment)

a. Lightweight anti-vibration chainsaw, of an approved pattern and suitable for chainsaw snedding, together with spares and maintenance tools.

b. Fuel and oil cans.

c. Breaking bar where required.

d. Logger's harness, spring loaded logger's tape (when primary conversion or measurement at stump is required).

e. Stump treatment equipment.

f. Protective clothing viz:
Safety helmet BS 2826, 2095 or 5240.
Eye protection.
Ear defenders.
Gloves, incorporating ballistic nylon lining on back of left hand.
Nylon leg guard.
Safety boots, incorporating inner ballistic nylon guard.

4 Allowances

The following allowances are included in the Output Guide:

a. For contingencies and work other than that actually performed on individual trees, eg refuelling and day-to-day maintenance of the saw, clearing lop and top from roads and rides, fetching tools and supplies etc 31% of the time actually spent on felling and snedding (table 6A).

Where the use of an extraction machine, carrying out primary extraction, causes unavoidable delays to the fellers, the 'other work' allowance is 45% of the time actually spent on felling and snedding (table 6B).

b. For personal needs and rest, 25% of the total working time (table 6A).

When the 'other work' allowance increases to 45%, personal needs and rest allowance is decreased to 23%.

N.B. (i) Any delays in excess of 11 minutes duration caused by the use of an extraction machine are not included and should be allowed for separately.

    (ii) Mechanical breakdowns of the saw in excess of 15 minutes are not included and should be allowed for separately.

    (iii) Times for primary or complete extraction are *excluded*.

c. Payment for worker owned chainsaws should be based on chainsaw utilisation of 70% when table 6A is used and 65% for table 6B.

5 Method of using the Output Guide

a. Select the appropriate table.

Use table 6A—Where complete snedding in situ is possible in safety (Working Method 1) or where the use of an extraction machine (Working Methods 2 and 3) does *not* cause delays in the felling and snedding operation.

Use table 6B—Where the use of an extraction machine (Working Methods 2 and 3) causes *unavoidable* delays in the felling and snedding operation.

b. Obtain mean tree volume by a recognised and locally agreed method.

c. Read the standard time per tree for the correct species from the appropriate table and modify as necessary with reference to paragraph 7.

Example 1

Sitka spruce. Mean tree volume $0.29\ m^3$.

Working Method 2. Extraction causing unavoidable delays in felling and snedding operation.

Brashing percentage 70%. Moderately difficult terrain.

Felling and snedding time (from table 6B).

Interpolated: $11.02 + \left[\dfrac{11.77 - 11.02}{4} \times 1 \right]$ = 11·21 SMs per tree

Allowance for moderately difficult terrain
(5%) (from paragraph 7a)   = 0·56 SMs per tree

Allowance for brashing (from paragraph 7b)
90% − 70% = 20% of 0·75 SMs   = 0·15 SMs per tree

Allowance for stump treatment
(from paragraph 7e)   = 0·60 SMs per tree

                     Total time = 12·52 SMs per tree

Example 2

Douglas fir. Mean tree volume 0·84 m³.

Working Method 3. Machine extraction does not cause delays in felling and snedding operation.

One multiple length sawlog measured, marked and crosscut from each tree.

Mean top diameter of sawlogs = 24 cm.

Conditions as shown in paragraph 1.

Felling and snedding time (from table 6A)   = 14·68 SMs per tree

Allowance for measure, mark and crosscut sawlog (from paragraph 7d).

interpolated: $0·70 + \left[\dfrac{0·79 - 0·70}{5} \times 4 \right]$ = 0·77 SMs per tree

Allowance for stump treatment (from paragraph 7e)   = 0·80 SMs per tree

Total time = 16·25 SMs per tree

6A Standard times for clearing windblown trees

Trees completely snedded in situ where possible in safety, or if primary extracted no unavoidable delay occurring.

Volume of average tree in m³	Standard time per tree in standard minutes		
	Sitka spruce	Norway spruce	Douglas fir
0·08	6·55	6·42	7·19
0·12	7·29	7·06	7·46
0·16	8·01	7·71	7·75
0·20	8·73	8·33	8·09
0·24	9·43	8·95	8·47
0·28	10·12	9·56	8·89
0·32	10·81	10·15	9·32
0·36	11·48	10·73	9·75
0·40	12·13	11·29	10·20
0·44	12·79	11·84	10·65
0·48	13·41	12·36	11·10
0·52	14·03	12·88	11·54
0·56	14·64	13·38	11·97
0·60	15·23	13·86	12·42
0·64	15·80	14·32	12·82
0·68	16·34	14·74	13·22
0·72	16·88	15·15	13·60
0·76	17·41	15·55	13·98
0·80	17·91	15·92	14·36
0·84	18·39	16·26	14·68
0·88	18·86	16·60	15·02
0·92	19·31	16·90	15·34
0·96	19·72	17·17	15·63
1·00	20·12	17·44	15·95
1·04	20·50	17·66	16·26
1·08	20·86	17·88	16·57
1·12	21·19	18·06	16·86
1·16	21·50	18·21	17·13
1·20	21·80	18·36	17·41

Interpolate as necessary.

6B Standard times for clearing windblown trees

Trees primary extracted whole or in part before snedding completed. The use of the extraction machine causing unavoidable delays to the felling and snedding operation.

Volume of average tree in m^3	Standard time per tree in standard minutes		
	Sitka spruce	Norway spruce	Douglas fir
0·08	7·13	6·99	7·83
0·12	7·94	7·69	8·13
0·16	8·72	8·40	8·44
0·20	9·51	9·07	8·81
0·24	10·27	9·75	9·23
0·28	11·02	10·41	9·69
0·32	11·77	11·05	10·15
0·36	12·50	11·69	10·62
0·40	13·22	12·30	11·11
0·44	13·93	12·90	11·60
0·48	14·61	13·46	12·09
0·52	15·28	14·03	12·57
0·56	15·94	14·57	13·04
0·60	16·59	15·10	13·53
0·64	17·21	15·60	13·96
0·68	17·80	16·05	14·40
0·72	18·39	16·50	14·81
0·76	18·96	16·94	15·23
0·80	19·51	17·34	15·64
0·84	20·03	17·71	15·99
0·88	20·55	18·08	16·36
0·92	21·03	18·41	16·71
0·96	21·47	18·70	17·02
1·00	21·92	18·99	17·37
1·04	22·33	19·23	17·71
1·08	22·72	19·47	18·05
1·12	23·08	19·67	18·36
1·16	23·42	19·83	18·66
1·20	23·74	20·00	18·96

Interpolate as necessary.

N.B. Where trees are partially converted at stump, extra time should be allowed for marking, measuring and crosscutting the sawlogs—see paragraph 7d.

7 Modifications and variations to the Output Guide

N.B. These modifications and variations are only to be applied when the prevailing conditions or job specification differ from those listed in paragraphs 1 and 2.

a. *Ground conditions*

Where ground conditions are difficult because of rocks, deep drains or excessive steepness

Add up to 10% in 5% steps.

For extremely difficult conditions

Add 15%.

b. *Brashing percentage*

If the percentage of trees brashed is less than 90% the brashing of all measurable unbrashed trees should be allowed for. The time for chainsaw brashing (or minimal brashing and extra snedding) is 0·75 SMs per tree.

c. *Measure and mark sawlog lengths*

When sawlog lengths are measured and marked (but *not* crosscut), during snedding using a logger's tape for measuring and the chainsaw for marking the trees

Add 0·45 standard minutes per tree for each timber length measured and marked.

d. *Measure, mark and crosscut sawlog length*

When a single or multiple sawlog length is measured, marked and crosscut during snedding make an addition from the following table:

Mean top diameter sawlogs	Time per sawlog
15 cm	0·64 SMs
20 cm	0·70 SMs
25 cm	0·79 SMs
30 cm	0·94 SMs
35 cm	1·16 SMs

e. *Stump treatment*

For applying fungicide to the stump immediately after felling (including fetching chemicals, filling cans, etc)

Add 0·40 SMs per tree for trees up to 0·18 m³

Add 0·60 SMs per tree for trees from 0·18 m³ to 0·72 m³

Add 0·80 SMs per tree for trees over 0·72 m³.

Recommended methods of working

Reference should be made to Leaflet 75, Harvesting of Wind-blown Trees, for further discussion on method and job organisation and safety aspects.

Methods

1 (i) Cut off tree at stump ensuring stump is made safe.
  (ii) Sned tree completely in situ.
  (iii) Treat stump, trim butt, cut off top, etc as in normal clear-felling.

2 (i) Cut off tree at stump ensuring stump is made safe.
  (ii) Sned tree to safe limit in situ.
  (iii) Primary extract tree clear of windblown tangle (about one tree length at least).
  (iv) Complete snedding.
  (v) Treat stump, trim butt, cut off top, etc as in normal clear-felling.

3 (i) Cut off tree at stump, ensuring stump is made safe.
  (ii) Sned tree to safe limit in situ.
  (iii) Measure and crosscut sawlog length(s) in situ.
  (iv) Extract sawlog(s) to roadside and extract remaining part of tree clear of windblown tangle (about one tree length at least).
  (v) Complete snedding.
  (vi) Treat stump, trim butt, cut off top, etc as in normal clear-felling.

Notes:

(a) Safe limit for snedding varies from tree to tree.

(b) Methods 2 and 3 apply in principle to both tractor and skyline extraction though the organisation of the latter is more complex.

(c) Methods 2 and 3 may involve some unavoidable delay to the feller in so far as he may have to wait for the extraction machine to remove the tree or part tree from the tangle, or the removal of a second tree may interfere with him as he works on a first tree. Delays are *not* inevitable when using these methods. They will depend on the nature of the blow, the type of tree, extraction method, etc.

1 **Conditions**

The standard times apply to:

a. Conversion of recently felled conifer poles into billets where there is a tolerance on length (eg pulpwood etc) and sawlogs if these have not already been cut at stump.

b. Poles in stacks of up to 100 poles.

c. Majority of poles lying the same way (ie butts together but not necessarily lined up).

d. Poles to be in an acceptable condition to meet the specifications, ie:
   (i) Root spurs taken off.
   (ii) Clean snedding with no brashing snags.

e. Poles parallel with road.

f. One-man working.

2 **Job specification**

The standard times are for the following work:

a. (i) Measuring and removal of one or more sawlogs and cutting the remaining piece into billets.
   (ii) Cutting into billets the remainder of a pole after a sawlog has been cut at stump.

b. All cuts to be made squarely across the pole.

c. Tops and other waste to be disposed of as directed by the supervisor.

d. *1·2m pulpwood:*
   (i) Dimensions:
      Bar setting: 1·15m.
      Length: 1·2m with tolerance down to 1·1m.
      Diameter: 9cm minimum UB—41cm maximum butt UB.
      Not more than 5cm taper on any one billet.
   (ii) Stacking:
      Complete stacking as directed by the supervisor.

e. *2·2 m pulpwood:*
   (i) Dimensions:
   Bar setting: 1·1 m.
   N.B. The method is to make a nick at 1·1 m and cut the
   billet at twice this, ie at 2·2 m to meet the length specification
   of 2·3 m with a tolerance down to 1·8 m.
   Diameter: 6·5 cm minimum UB—40 cm maximum butt OB.
   (ii) Stacking:
   Billets to be stacked on bearers with ends flush together and
   at right angles to the road for subsequent loading by
   hydraulic grapple.

f. *Timber:*
   (i) Dimensions:
   Timber to be cut off at 14 cm top diameter UB.
   (ii) Stacking:
   Logs to be rolled to side of road.

3 Tools and equipment (including safety equipment)

a. Chainsaw with appropriate guidebar length—where the measuring bar attachment takes up some of the length of the blade, the guidebar length should be:
   Sawlogs 14 cm diameter maximum: guidebar at least 30 cm.
   Sawlogs above 14 cm diameter: guidebar at least 40 cm.

b. Spares, tool kit and fuel cans for chainsaw.

c. Measuring bar of a type that swings either side of saw.

d. Protective clothing viz:
   Safety helmet BS 2826, 2095 or 5240.
   Eye protection.
   Ear defenders.
   Gloves, incorporating ballistic nylon lining on back of left hand.
   Nylon leg guard.
   Safety boots, incorporating inner ballistic nylon guard.

e. Spring loaded tape and harness for measuring timber.

f. 1 m long handled hookeroon for stacking 2·2 m billets.

g. Spring loaded hand tongs or pulphook for stacking 1·2 m billets.

4 Allowances

Included in the standard time

a. Contingencies and other work, eg general preparation, start, refuel and maintain chainsaw, fixing and checking the measuring bar, minor trimming of poles etc, 20% of the time spent on crosscutting and stacking.

b. Personal needs and rest:
2·2 m billets 22–24% of the total working time, according to pole size.
1·2 m billets 24–32% of the total working time, according to pole size.

To be included in the price per standard minute
c. The appropriate incentive allowance.

5 Method of selecting the standard time
a. *Where all conversion is at roadside*
One of the following methods:
(i) The normal tariff of the compartment (see FC Booklet No 36) when using the standard time table based on volume (table 6A).
(ii) A random sample of not less than 30 poles from roadside stacks, measured to 7 cm top diameter, when using the standard time table based on length, *or* where no assessment of stand volume has been made (table 6B).

b. *Where logs have been removed at stump*
One of the following methods:
(i) A random sample of not less than 30 poles from roadside stacks, measured to 7 cm top diameter.
(ii) Measure the sawlogs cut at stump, deduct from the total stand volume, and divide the remaining volume by the number of poles to be crosscut.

6A Standard times for crosscutting poles at roadside using chainsaw and measuring bar

By volume

Mean pole volume (m^3)	1·2 m billets	2·2 m billets
	SMs per pole to measure, crosscut, stack	
·06	1·8	2·0
·07	2·0	2·1
·08	2·2	2·2
·09	2·4	2·3
·10	2·6	2·4
·11	2·8	2·6
·12	2·9	2·7
·13	3·1	2·8
·14	3·2	2·9
·15	3·4	3·0
·16	3·5	3·1
·17	3·7	3·2

Mean pole volume (m³)	1·2m billets	2·2m billets
	SMs per pole to measure, crosscut, stack	
·18	3·8	3·3
·19	4·0	3·4
·20	4·1	3·5
·22	4·3	3·7
·24	4·5	3·8
·26	—	3·9
·28	—	4·0
·30	—	4·1
·32	—	4·2
·36	—	4·3
·44	—	4·5

6B Standard times for crosscutting poles at roadside using chainsaw and measuring bar

By length

Mean pole length (m)	1·2m billets	2·2m billets
	SMs per pole to measure, crosscut, stack	
6·0	2·2	—
6·5	2·4	—
7·0	2·6	2·4
7·5	2·8	2·6
8·0	3·0	2·8
8·5	3·2	3·0
9·0	3·4	3·2
9·5	3·6	3·4
10·0	3·8	3·6
10·5	4·0	3·7
11·0	4·2	3·8
12·0	—	4·0
13·0	—	4·1
14·0	—	4·2
16·0	—	4·3

7 Modifications and variations to the standard times

N.B. These modifications and variations are only to be applied when the prevailing conditions or job specification differ from those listed in paragraphs 1 and 2.

a. *Excessively dirty poles*
When the extracted poles are excessively muddy resulting in extra sharpening of the saw's chain
Add up to 10% (in steps of 5%)
It should be stressed that frequent light sharpening (taking about 3 minutes) is preferable to a lengthy operation at longer intervals.

b. *Distance between stacks*
Included in the standard times is an allowance to cater for walking between stacks 25 metres apart. Where stacks are in excess of this
Add 5% per 100 metres

c. *Size of stacks (2·2m billets)*
Where the number of poles per stack (presented parallel to road) exceeds 100
Add up to 15% (in steps of 5%)

d. *Mark/book timber*
For each piece of timber marked
Add 0·8 standard minutes
For each piece of timber marked and entered in a book
Add 1·1 standard minutes

Standard Times for Chainsaw Crosscutting at Roadside (Measuring With a Rod and Scribe)
South Wales

1 Conditions

The standard times apply to the following conditions:

a. Crosscutting conifer species at roadside.

b. Poles in an acceptable condition to meet the product specification, ie root spurs taken off and clean snedded with only occasional stubs or branches to be removed during crosscutting (but see paragraph 7b).

c. Poles stacked parallel to the road on bearers.

d. Stacks of up to approximately 120 poles for poles of less than $0.2\,m^3$ average volume and up to 30 poles per stack for poles of $0.2\,m^3$ and over.

e. One-man working (see Appendix I for working method).

2 Job Specification

The standard times are for the following work:

a. Product mixes are combinations of the following:
   (i)   Millwood of set lengths with a minimum top diameter of 16 cm under bark.
   (ii) Chipper logs—lengths 1·85 m, 2·15 m, 2·46 m, 2·76 m, 3·07 m with a top diameter under bark of 13–18 cm.
   (iii) Pitwood cut to South Wales sizes.

   Note: 1·2 m pulpwood is not included in the tables in paragraph 6, but for recommendations see paragraph 7f.

b. All sawcuts to be made squarely across the pole.

c. All produce, except millwood, to be stacked suitably for mechanical loading. (If millwood needs to be moved see paragraph 7g.)

d. Roads and drains to be left clear of all brash and waste.

255

3 **Tools and equipment** (including safety equipment)
   a. Lightweight AV chainsaw (with a cubic capacity of approximately 68 cc) with maintenance tools.

   Note: If a lower capacity saw is used see paragraph 7h.

   b. Fuel cans.

   c. Rule for measuring top diameters.

   d. Measuring rod and scribe.

   e. Felt markers or crayons.

   f. One or two pulp hooks (for stacking pitwood and pulpwood) and one timber tongs (for moving larger produce).

   g. Breaking bar with cant-hook attachment (for moving millwood during measure and mark).

   h. Protective clothing viz:
   Safety helmet BS 2826, 2095 or 5240.
   Eye protection.
   Ear defenders.
   Gloves, incorporating ballistic nylon lining on back of left hand.
   Nylon leg guard.
   Safety boots, incorporating inner ballistic nylon guard.

4 **Allowances**

   Included in the standard times

   a. Contingencies and other work:
   (i) For contingencies and work other than walk and prepare, measure and crosscut, eg tool maintenance, moving poles, walking between stacks etc, 54% of the time spent on walk and prepare, measure and crosscut.
   (ii) For contingencies and work other than stacking, eg tidy stacks, walk between stacks etc, 14% of the time spent on stacking.

   b. Personal needs and rest:
   (i) For walk and prepare, measure and crosscut, 23% of the total working time for these operations.
   (ii) For stacking, from 19% to 49% of the total stacking time according to the volumes of the pieces stacked.

   To be included in the price per standard minute
   c. The appropriate incentive allowance.

5 **Method of selecting the standard time**
   a. *Time per pole—tables 6(a)(i) and 6(a)(ii)*
   Table 6(a)(i) gives standard times for walk and prepare, measure and crosscut per pole. To obtain the standard time, the mean volume and the mean number of pieces cut per pole are required.

The mean volume will normally be the tariff volume (see FC Booklet No 36) unless primary conversion of millwood has been done at stump. For guidance in determining millwood content see Appendix III.

The mean number of pieces cut per pole can only be calculated accurately at the end of the job by dividing total number of pieces by total number of poles. Alternatively samples can be taken during working but for accuracy they need to be spread through the crop.

It is suggested that, unless the supervisor's experience suggests otherwise, the mean figure in column 3 may be taken as a starting point and adjustment can be made to the standard times as information on the number of pieces is gathered.

Note: If the mean number of pieces per pole differs from that in column 3 of the table *add* or *subtract* 0·045 standard minutes for each tenth of a piece different from the tabulated mean number of pieces.

Table 6(a)(ii) gives standard times for stacking.

To obtain the standard time for stacking per pole, mean number of pieces stacked per pole, and mean volume of the pieces stacked are required.

Where no millwood is produced this simply means dividing the original pole volume by the number of pieces cut per pole. The standard time for stacking appropriate to this mean volume per piece is then selected from table 6(a)(ii); the standard time per pole is obtained by multiplying this figure by the mean number of pieces per pole (see example 1 in Appendix II).

Where millwood is produced the mean pole volume must be reduced by the total volume of millwood cut per pole before table 6(a)(ii) is used and the number of pieces must be reduced by the number of millwood pieces cut per pole. (See example 2 in Appendix II.)

b. *Time per piece—tables 6(b)(i), 6(b)(ii) and 6(b)(iii)*

   Table 6(b)(i)   Standard time per piece by volume for walk and prepare, measure, crosscut and stack.
   The underbark volume per piece in each size of produce (excepting millwood) is required.

   Table 6(b)(ii)   Standard time per piece by length and top diameter (UB) for walk and prepare, measure, crosscut and stack.
   Lengths and top diameters (UB) of pitwood and chipper log pieces are required.

   Table 6(b)(iii)   Standard time per millwood piece for walk and prepare, measure and crosscut.
   Lengths and top diameters (UB) of millwood pieces are required.

6A(i) Standard time per pole for walk and prepare, measure and crosscut

Volume of pole in m³ (OB)	Standard time in standard minutes	Mean number of pieces per pole
0·02	1·81	4·6
0·04	2·00	4·9
0·06	2·23	5·1
0·08	2·47	5·4
0·10	2·69	5·7
0·12	2·89	5·9
0·14	3·09	6·1
0·16	3·29	6·3
0·18	3·50	6·6
0·20	3·70	6·8
0·22	3·90	6·9
0·24	4·10	7·1
0·26	4·30	7·3
0·28	4·49	7·4
0·30	4·67	7·6
0·32	4·85	7·7
0·34	5·04	7·8
0·36	5·22	8·0
0·38	5·39	8·1
0·40	5·57	8·2
0·42	5·74	8·3
0·44	5·90	8·4
0·46	6·06	8·5
0·48	6·22	8·5
0·50	6·37	8·6
0·54	6·68	8·7
0·58	6·98	8·8
0·62	7·24	8·9
0·66	7·51	9·0
0·70	7·77	9·0
0·74	7·96	9·0
0·78	8·16	9·0
0·82	8·35	8·9
0·86	8·51	8·8
0·90	8·65	8·7
0·94	8·78	8·6
0·98	8·88	8·5

Add or *subtract* 0·045 standard minutes for each tenth of a piece more or less than the mean number of pieces shown in column 3.

6A(ii) Standard time per mean piece for stacking

Piece volume in m³ (OB)	Standard time in standard minutes
0·008	0·27
0·010	0·28
0·012	0·29
0·014	0·29
0·016	0·30
0·018	0·31
0·020	0·32
0·022	0·33
0·024	0·34
0·026	0·35
0·028	0·36
0·030	0·38
0·032	0·39
0·034	0·40
0·036	0·41
0·038	0·43
0·040	0·45
0·042	0·46
0·044	0·48
0·046	0·49
0·048	0·51
0·050	0·52

Standard time per piece by volume for crosscutting at roadside

6B(i) *Standard time per piece for walk and prepare, measure, crosscut and stack*

Piece volume in m³ (UB)	Standard time in standard minutes
0·004	0·70
0·006	0·72
0·008	0·74
0·010	0·76
0·012	0·79
0·014	0·81
0·016	0·84
0·018	0·86
0·020	0·88
0·022	0·91
0·024	0·93
0·026	0·95
0·028	0·98
0·030	1·01
0·032	1·03
0·034	1·06
0·036	1·08
0·038	1·10
0·040	1·13
0·042	1·16
0·044	1·18
0·046	1·21
0·048	1·23
0·050	1·26
0·052	1·29
0·054	1·31
0·058	1·36
0·062	1·42
0·066	1·48
0·070	1·54
0·074	1·60
0·078	1·66
0·082	1·73

Standard time per piece for crosscutting at roadside

6B(ii) *Standard time per piece for walk and prepare, measure,*
crosscut and stack

Since pitwood is a predetermined specification the following
standard time per piece by pitwood specification is given (derived
from table 6(b)(i)).

Pitwood size in mm	Volume in m³	Time per piece in standard minutes	Pitwood size in mm	Volume in m³	Time per piece in standard minutes
900 × 70	0·004	0·70	1,500 × 130	0·023	0·92
1,050 × 70	0·005	0·71	1,500 × 140	0·026	0·95
1,050 × 80	0·006	0·72	1,500 × 150	0·030	1·01
1,050 × 90	0·008	0·74	1,500 × 160	0·034	1·06
1,050 × 100	0·009	0·75	1,500 × 170	0·038	1·10
1,050 × 110	0·011	0·77	1,650 × 130	0·025	0·94
1,200 × 90	0·009	0·75	1,800 × 150	0·037	1·09
1,200 × 100	0·011	0·77	1,800 × 160	0·041	1·14
1,200 × 130	0·018	0·86	1,950 × 140	0·035	1·07
1,350 × 70	0·006	0·72	1,950 × 150	0·040	1·13
1,350 × 80	0·008	0·74	1,950 × 160	0·045	1·19
1,350 × 90	0·010	0·76	1,950 × 170	0·051	1·27
1,350 × 100	0·012	0·79	2,250 × 150	0·047	1·22
1,350 × 110	0·015	0·82	2,700 × 150	0·058	1·36
1,350 × 120	0·017	0·85	2,700 × 160	0·066	1·48
1,350 × 130	0·020	0·88	2,700 × 170	0·073	1·58
1,500 × 120	0·020	0·88	2,700 × 180	0·082	1·73

Standard time per piece for chipper logs

Length in m	Top diameter (UB) in cm					
	13	14	15	16	17	18
1·85	1·07	1·13	1·15	1·19	1·29	1·39
2·15	1·13	1·19	1·23	1·32	1·42	1·54
2·46	1·19	1·26	1·32	1·39	1·54	1·69
2·76	1·26	1·39	1·46	1·54	1·69	
3·07	1·32	1·45	1·54	1·69		

6B(iii) *Standard time per piece for crosscutting millwood at roadside*

Length in m	Top diameter (UB) in cm														
	16	18	20	22	24	26	28	30	32	34	36	38	40	42	44
1·8	0·52	0·56	0·62	0·71	0·78	0·86	0·97	1·07	1·16	1·26	1·39	1·52	1·65	1·80	1·94
2·0	0·57	0·61	0·68	0·76	0·83	0·91	1·02	1·12	1·21	1·31	1·44	1·57	1·70	1·85	1·99
2·2	0·62	0·67	0·73	0·81	0·88	0·96	1·07	1·17	1·27	1·37	1·49	1·63	1·76	1·90	2·04
2·4	0·67	0·72	0·78	0·86	0·93	1·01	1·12	1·22	1·32	1·42	1·54	1·68	1·81	1·96	2·09
2·6	0·72	0·77	0·84	0·92	0·99	1·07	1·18	1·28	1·37	1·47	1·60	1·73	1·87	2·02	2·15
2·8	0·77	0·82	0·90	0·97	1·05	1·12	1·23	1·33	1·43	1·52	1·65	1·78	1·92	2·07	2·20
3·0	0·83	0·87	0·95	1·02	1·10	1·17	1·29	1·38	1·48	1·57	1·70	1·83	1·97	2·12	2·25
3·2	0·88	0·91	0·99	1·07	1·15	1·22	1·33	1·43	1·52	1·62	1·75	1·88	2·01	2·16	2·30
3·4	0·93	0·96	1·04	1·12	1·19	1·27	1·38	1·48	1·57	1·67	1·80	1·93	2·06	2·21	2·35
3·6	0·98	1·02	1·09	1·17	1·24	1·32	1·43	1·53	1·62	1·72	1·85	1·98	2·11	2·26	2·40
3·8	1·03	1·07	1·14	1·22	1·29	1·37	1·48	1·58	1·67	1·77	1·90	2·03	2·16	2·31	2·45
4·0	1·08	1·12	1·19	1·27	1·34	1·42	1·53	1·63	1·72	1·82	1·95	2·08	2·21	2·36	2·50
4·2	1·12	1·16	1·24	1·31	1·39	1·46	1·58	1·67	1·77	1·86	1·99	2·13	2·26	2·41	2·54
4·4	1·17	1·21	1·29	1·36	1·44	1·51	1·63	1·72	1·82	1·91	2·04	2·18	2·31	2·46	2·59
4·6	1·22	1·26	1·34	1·41	1·49	1·56	1·68	1·77	1·87	1·96	2·09	2·23	2·36	2·51	2·64
4·8	1·27	1·31	1·39	1·46	1·54	1·61	1·73	1·82	1·92	2·01	2·14	2·28	2·41	2·56	2·69
5·0	1·32	1·36	1·44	1·51	1·59	1·66	1·78	1·87	1·97	2·06	2·19	2·33	2·46	2·61	2·74

7 Modifications and variations to the standard times

N.B. These modifications and variations are only to be applied when the prevailing conditions or job specification differ from those listed in paragraphs 1 and 2.

a. *Dirty poles*

When poles are excessively gritty resulting in extra sharpening of the chain

Add 5% to the total standard time per pole or piece when using tables 6(a)(i), 6(b)(i) and 6(b)(ii)

Add 10% to the total standard time per piece when using tables 6(b)(iii)

b. *Extra snedding*

Where extra snedding is necessary during crosscutting

Add 5% to the total standard time per pole or piece when using tables 6(a)(i), 6(b)(i) and 6(b)(ii)

Add 10% to the total standard time per piece when using table 6(b)(iii)

(Extra snedding is usually required where large poles are deliberately left unturned at the felling and snedding stage.)

c. *Taking up produce (other than millwood)*

(i) When using table 6(a)(ii)

For counting and recording the numbers of pieces of each specification of produce

Add 12½% to the standard time for stacking.

(ii) When using table 6(b)(i) or 6(b)(ii)

For counting and recording the numbers of pieces of each specification of produce

Add 5% to the total standard time per piece.

d. *Measuring, marking and booking millwood pieces*

For measuring top diameters (UB), marking these top diameters on the log with a crayon or felt marker and entering the information in a book or on a form as required

Add 0·50 standard minutes per piece.

e. *Restricted stacking*

Normally when crosscutting a pile of poles the *average* distance produce is moved for stacking is not more than approximately 3 metres. If on average produce has to be moved further, additions up to those indicated below are suggested.

(i) When using either table 6(b)(i) or 6(b)(ii)

Add 5% for pieces up to 0·02 cubic metres

Add 10% for pieces from 0·02 cubic metres to 0·05 cubic metres

Add 20% for pieces greater than 0·05 cubic metres

(ii) When using table 6(a)(ii)

Add 15% if the mean piece is below 0·02 cubic metres

> **Add** 20% if the mean piece is between 0·02 cubic metres and
> 0·03 cubic metres
> **Add** 25% if the mean piece is greater than 0·03 cubic metres

f. *Production of 1·2 m pulpwood*

The following recommendation applies only to pulpwood (length 1·2 metres to 1·1 metres with a top diameter range of 9 to 12 centimetres under bark) cut in conjunction with chipper logs and pitwood, or millwood, chipper logs and pitwood.

For each pulpwood billet cut

Subtract 0·15 standard minutes

Note: Where the object is maximum production of 1·2 m pulpwood, reference should be made to the table for crosscutting using a measuring bar (XIX/21).

g. *Minimum movement of millwood pieces*

If it is necessary to move millwood pieces by hand so that traffic is not impeded

> **Add** 0·20 standard minutes per millwood piece moved (when movement of millwood is by rolling it to the side of the road).

h. *Use of saws of less than 68 cc*

If a lower powered saw such as a Jonsered 621 (56 cc) or Husqvarna 160S (58 cc) is used for crosscutting, make the following additions to the standard times:

(i) When using table 6(b)(iii) add to the standard time per piece as indicated below:

Diameter of saw cut in centimetres	*Addition in standard minutes per piece*
Less than 25	0
25–29	0·05
30–34	0·07
35–39	0·10
40 and over	0·12

(ii) When using table 6(a)(i)

> **Add** 5% if the mean pole is between 0·40 cubic metres and 0·80 cubic metres
> **Add** 10% if the mean pole is greater than 0·80 cubic metres

APPENDIX I

Presentation of poles for crosscutting and recommended working method

Presentation of poles for crosscutting

1. Poles parallel to the road.
2. Poles on bearers.
3. Presentation must be all tips or all butts together so that cut

pieces lie approximately opposite stacked pieces of the same size.

4. Where extraction and crosscutting are carried out simultaneously a grid may be constructed. Studies indicate a saving in crosscutting time when a grid is used.

Recommended working method (one-man working)

1. Pick up measuring rod and scribe.
2. Walk to stack.
3. Measure from butt end and scribe as many pieces as possible in the stack.
4. Aside measuring rod and scribe.
5. Pick up saw.
6. Walk to stack.
7. Crosscutt scribed pieces.
8. Put aside saw.
9. Stack crosscut pieces.

   Repeat sequence until stack is completed.

Note:

   (i) Stacking of millwood pieces is not included in the standard times.
   (ii) Pieces in excess of 0·05 m³ in volume should be rolled or lifted one end at a time on to the stack.

APPENDIX II

Method of selecting the standard time

Example 1

Average pole from tariff = 0·14 m³
Mean number of pieces cut per pole = 5·3

Standard time for measure and crosscut per pole
= Standard time from 6(a)(i) for 0·14 m³ minus adjustment for number of pieces cut
= 3·09 standard minutes − [0·045 × (6·1 − 5·3)] standard minutes
= 3·09 standard minutes − (0·045 × 8) standard minutes
= 3·09 − 0·36 standard minutes
= 2·73 standard minutes

Standard time for stacking (for a pole from which no timber is produced)
Mean volume per piece stacked

$$= \frac{\text{average pole volume from tariff}}{\text{mean number of pieces cut}}$$

$$= \frac{0·14 \, \text{m}^3}{5·3} = 0·026 \, \text{m}^3$$

265

Standard time for stacking per piece from table 6(a)(ii)
$=0{\cdot}35$ standard minutes
Standard time for stacking per pole
$=0{\cdot}35\times5{\cdot}3$ standard minutes
$=1{\cdot}85$ standard minutes
Total standard time per pole $=$ standard time for measure and cross-cut plus standard time for stacking
$=2{\cdot}73+1{\cdot}85$
Total standard time per pole $=4{\cdot}58$ standard minutes

Example 2
Average pole volume from tariff $=0{\cdot}59\,\mathrm{m}^3$
Mean number of pieces cut per pole (assessed) $=9{\cdot}9$ pieces
Total volume of millwood per pole $=0{\cdot}45\,\mathrm{m}^3$
Total number of pieces of millwood per pole $=4{\cdot}6$ pieces

Standard time for measure and crosscut
$=$ Standard time from 6(a)(i) for $0{\cdot}59\,\mathrm{m}^3$ plus adjustment for number of pieces cut
$=7{\cdot}03+(0{\cdot}045\times10)$ standard minutes
$=7{\cdot}03+0{\cdot}45$ standard minutes
$=7{\cdot}48$ standard minutes

Standard time for stacking
Mean volume per piece stacked
$$=\frac{\text{mean pole volume minus millwood volume}}{\text{assessed number of pieces minus millwood pieces per pole}}$$
$$=\frac{0{\cdot}59-0{\cdot}45}{9{\cdot}9\;-4{\cdot}6}=\frac{0{\cdot}14}{5{\cdot}3}=0{\cdot}026$$
Standard time for stacking per piece from table 6(a)(ii)
$=0{\cdot}35$ standard minutes
Standard time for stacking per pole
$=0{\cdot}35\times5{\cdot}3$ standard minutes
$=1{\cdot}85$ standard minutes
Total standard time per pole $=$ standard time for measure and cross-cut plus standard time for stacking
$=7{\cdot}48+1{\cdot}85$ standard minutes
$=9{\cdot}33$ standard minutes

APPENDIX III

A method of assessing timber volume and number of pieces of millwood per tree

1. Tables 1 and 2 are tree assortment tables giving length to top diameter underbark of 16 and 18 cm respectively for trees of various breast height diameter classes and by height classes

(height class 18 means trees with total heights between $16\frac{1}{2}$ and $19\frac{1}{2}$ metres etc).

2. Tables 3 and 4 show the volume in cubic metres by millwood length and tree height class for top diameters underbark of 16 and 18 cm respectively.

   Notes:
   (i) Tables 1, 2, 3 and 4. The top diameter underbark is the diameter to which millwood is cut.
   (ii) Tables 1, 2, 3 and 4. The height class refers to the overall height of the tree. The most accurate method of finding the height of each size of tree is by making comparison with the sample trees on the U15 (D) using lengths to 7 cm which are shown in brackets under the height classes on the tables. Alternatively, assess the total height of the trees to be felled and drop a class to obtain a figure for the mean tree which is quicker but less accurate.
   (iii) Tables 1 and 2. The breast height diameter class is taken from the girthing record (C) on the U15 (tariff sheet).

3. *Method of working*
   (i) For each size of tree on the girthing record on the U15 (C) read the length to the required underbark top diameter from the appropriate table (1 or 2) against the appropriate breast height diameter and tree height class.
   (ii) From the length so obtained assess the number and lengths of pieces of millwood that can be cut from each size of tree and add their lengths together to obtain the length of timber to be obtained from the tree, which may be the same as, or less than the length obtained from the table.
   (iii) Using the length of millwood in each size of tree obtain its volume from table 3 or 4 as appropriate, using the same tree height class (or height to 7 cm) as in paragraph 1.
   (iv) For each breast height diameter size of tree multiply the number of pieces of timber and the volume of timber in the tree by the number of trees of that size in the stand. Sum the total number of pieces and the total volume of timber in all the trees to be felled and divide each total by the number of trees to obtain the mean number of pieces and the mean volume of timber per tree.

   Example
   Millwood specifications
   1. 1·85 metres in length with a top diameter of 16 cm underbark or more.
   2. 2·65 metres in length with a top diameter of 16 cm underbark or more.

Breast height dia cm	Number of trees	Height class in metres	Lengths to 16 cm under-bark in metres	Esti-mated numbers of pieces per pole	Total length of esti-mated pieces in m	Vol. of esti-mated pieces in pole (m³)	Total number of pieces in BH dia class (2×5)	Total volume in BH dia class (m³) (2×7)
(1)	(2)	(3)	(4)	(5)	(6)	(7)	(8)	(9)
20	5	18	5·3	2(2 × 2·65)	5·3	0·149	10	0·745
21	—	—	—	—	—	—	—	—
22	7	18	7·1	3(1 × 2·65) (2 × 1·85)	6·4	0·182	21	1·274
23	8	18	7·9	3(2 × 2·65) (1 × 1·85)	7·2	0·215	24	1·720
24	10	18	8·5	4(1 × 2·65) (3 × 1·85)	8·2	0·272	40	2·720
25	14	18	9·1	4(2 × 2·65) (2 × 1·85)	9·0	0·325	56	4·550
26	20	18	9·6	4(2 × 2·65) (2 × 1·85)	9·0	0·325	80	6·500
27	22	18	10·0	4(3 × 2·65) (1 × 1·85)	9·8	0·392	88	8·624
28	15	18	10·4	4(3 × 2·65) (1 × 1·85)	9·8	0·392	60	5·880
29	7	18	10·7	4(4 × 2·65)	10·6	0·472	28	3·304
30	6	18	11·0	4(4 × 2·65)	10·6	0·472	24	2·832
31	5	18	11·3	5(2 × 2·65) (3 × 1·85)	10·9	0·508	25	2·540
32	1	18	11·5	5(2 × 2·65) (3 × 1·85)	10·9	0·508	5	0·508
Total	120						461	41·197

$$\text{Total volume of millwood per tree} = \frac{\text{Total col 9}}{\text{Total col 2}} = \frac{41\cdot197}{120} = 0\cdot343$$

$$\text{Number of pieces of millwood per tree} = \frac{\text{Total col 8}}{\text{Total col 2}} = \frac{461}{120} = 3\cdot84$$

Notes:

Column 1 Taken directly from Girthing Record (U15(c)).

Column 2 Taken directly from Girthing Record (U15(c)).

Column 4 From table 1 (breast height diameter and appropriate height class).

Column 5 Probable utilisation of the length according to the millwood specifications (figures in brackets are an explanation of the number of pieces).

Column 6 The total length of the pieces in column 5.

Column 7 From table 3 (the appropriate volume for the length quoted in column 6 according to height class).

Column 8 Column 5 multiplied by column 2.

Column 9 Column 7 multiplied by column 2.

TABLE 1

General tree assortment table for conifers
Lengths to various top diameters underbark

Under-bark top dia (in cm)	Breast height dia class (in cm)	Height class in metres (height to 7 cm in brackets)									
		9 (6·5)	12 (9·5)	15 (12)	18 (15)	21 (18)	24 (21)	27 (24)	30 (27)	33 (30)	36 (33)
16	20	2·7	3·3	4·5	5·3	6·3	8·0				
	21	3·1	3·9	5·1	6·3	7·3	9·2				
	22	3·5	4·4	5·7	7·1	8·3	10·3				
	23	3·8	4·9	6·3	7·9	9·2	11·4				
	24	4·3	5·3	6·8	8·5	10·0	12·3				
	25	4·6	5·8	7·3	9·1	10·7	13·0				
	26		6·2	7·7	9·6	11·3	13·7				
	27		6·5	8·0	10·0	11·8	14·2				
	28		6·8	8·4	10·4	12·2	14·7				
	29		7·1	8·7	10·7	12·6	15·1				
	30		7·3	9·0	11·0	13·0	15·5	17·5	20·0		
	31			9·2	11·3	13·3	15·8	17·8	20·3		
	32			9·4	11·5	13·6	16·1	18·1	20·6		
	33			9·6	11·7	13·9	16·4	18·4	20·9		
	34			9·8	11·9	14·2	16·6	18·7	21·2		
	35			10·0	12·1	14·4	16·8	18·9	21·4		
	36			10·2	12·3	14·6	17·0	19·1	21·6		
	37			10·4	12·4	14·8	17·2	19·3	21·8		
	38			10·5	12·6	14·9	17·3	19·5	22·0		
	39			10·6	12·7	15·0	17·4	19·7	22·2		
	40			10·8	12·8	15·1	17·5	19·8	22·4	24·8	28·1
	41					15·2	17·6	20·0	22·5	25·0	28·3
	42					15·2	17·7	20·1	22·7	25·2	28·5
	43					15·3	17·8	20·2	22·8	25·4	28·6
	44					15·4	17·9	20·3	23·0	25·6	28·7
	45					15·4	18·0	20·4	23·1	25·8	28·8
	46					15·5	18·0	20·5	23·2	25·9	29·0
	47					15·5	18·1	20·6	23·3	26·0	29·1
	48					15·6	18·1	20·7	23·4	26·2	29·2
	49					15·6	18·2	20·8	23·5	26·3	29·3
	50					15·7	18·3	20·8	23·6	26·4	29·4
	51						18·3	20·9	23·7	26·5	29·5
	52						18·3	21·0	23·7	26·6	29·6
	53						18·4	21·0	23·8	26·7	29·7
	54						18·4	21·1	23·9	26·8	29·7
	55						18·5	21·1	24·0	26·9	29·8
	56							21·2	24·1	27·0	29·9
	57							21·2	24·1	27·0	30·0
	58							21·3	24·2	27·1	30·0
	59							21·4	24·3	27·2	30·1
	60							21·5	24·4	27·2	30·2

TABLE 2

General tree assortment table for conifers
Lengths to various top diameters underbark

Under-bark top dia (in cm)	Breast height dia class (in cm)	Height class in metres (height to 7 cm in brackets)									
		9 (6·5)	12 (9·5)	15 (12)	18 (15)	21 (18)	24 (21)	27 (24)	30 (27)	33 (30)	36 (33)
18	20		2·0	2·5	3·5	4·0	4·5				
	21	2·0	2·6	3·2	4·3	5·0	5·8				
	22	2·6	3·2	3·9	5·2	6·0	7·0				
	23	3·1	3·7	4·5	6·0	6·9	8·2				
	24	3·6	4·2	5·1	6·6	7·7	9·3				
	25	4·0	4·6	5·7	7·2	8·5	10·3				
	26		5·0	6·2	7·8	9·2	11·2				
	27		5·4	6·6	8·3	9·7	11·9				
	28		5·8	7·0	8·8	10·3	12·6				
	29		6·2	7·4	9·2	10·8	13·1				
	30		6·5	7·8	9·6	11·3	13·6	15·3	17·8		
	31			8·2	10·0	11·7	14·0	15·7	18·2		
	32			8·5	10·3	12·1	14·4	16·1	18·6		
	33			8·8	10·5	12·4	14·7	16·5	19·0		
	34			9·0	10·8	12·7	15·0	16·9	19·4		
	35			9·2	11·0	13·0	15·3	17·3	19·7		
	36			9·5	11·2	13·3	15·6	17·6	20·0		
	37			9·7	11·4	13·5	15·8	17·9	20·3		
	38			9·9	11·5	13·7	16·0	18·2	20·6		
	39			10·1	11·5	13·9	16·2	18·4	20·8		
	40			10·2	11·6	14·1	16·4	18·6	21·0	23·4	26·4
	41					14·3	16·6	18·8	21·2	23·7	26·6
	42					14·4	16·8	19·0	21·4	24·0	26·8
	43					14·5	16·9	19·2	21·6	24·2	27·0
	44					14·6	17·0	19·4	21·8	24·4	27·2
	45					14·7	17·1	19·5	22·0	24·6	27·4
	46					14·8	17·2	19·6	22·2	24·8	27·6
	47					14·9	17·3	19·7	22·4	25·0	27·8
	48					14·9	17·4	19·8	22·5	25·2	28·0
	49					15·0	17·5	19·9	22·6	25·4	28·2
	50					15·0	17·5	20·0	22·7	25·6	28·3
	51						17·6	20·1	22·8	25·7	28·4
	52						17·7	20·2	22·9	25·8	28·5
	53						17·7	20·3	23·0	25·9	28·6
	54						17·8	20·4	23·1	26·0	28·7
	55						17·8	20·5	23·2	26·1	28·8
	56							20·5	23·3	26·2	28·9
	57							20·6	23·4	26·3	29·0
	58							20·6	23·5	26·4	29·1
	59							20·7	23·6	26·4	29·2
	60							20·8	23·7	26·5	29·3

General tree assortment table for conifers
Volume (OB) to various top diameters underbark

TABLE 3

Underbark top dia (in cm)	Length in metres	Height class in metres (height to 7 cm in brackets)									
		9 (6·5)	12 (9·5)	15 (12)	18 (15)	21 (18)	24 (21)	27 (24)	30 (27)	33 (30)	36 (33)
16	2·0	0·055									
	2·5	0·080									
	3·0	0·100	0·085								
	3·5	0·120	0·105								
	4·0	0·150	0·125								
	4·5	0·180	0·150	0·130							
	5·0		0·180	0·145	0·140						
	5·5		0·205	0·170	0·155						
	6·0		0·240	0·195	0·170	0·165					
	6·5		0·280	0·220	0·185	0·180					
	7·0		0·320	0·250	0·205	0·200					
	7·5			0·285	0·230	0·220					
	8·0			0·330	0·260	0·240	0·225				
	8·5			0·380	0·290	0·260	0·235				
	9·0			0·430	0·325	0·280	0·250				
	9·5			0·500	0·365	0·305	0·270				
	10·0			0·580	0·410	0·335	0·290				
	10·5			0·690	0·460	0·365	0·310				
	11·0				0·520	0·400	0·330				
	11·5				0·590	0·440	0·355				
	12·0				0·680	0·485	0·385				
	12·5				0·790	0·535	0·415				
	13·0					0·595	0·450				
	13·5					0·660	0·490				
	14·0					0·740	0·535				
	14·5					0·835	0·585				
	15·0					0·990	0·640				
	15·5					1·280	0·710				
	16·0						0·785				
	16·5						0·875				
	17·0						1·000				
	17·5						1·180	0·790			
	18·0						1·460	0·850			
	18·5						2·000	0·940			
	19·0							1·040			
	19·5							1·190			
	20·0							1·390	0·905		
	20·5							1·700	0·980		
	21·0							2·120	1·060		
	21·5							2·690	1·175		
	22·0								1·310		
	22·5								1·500		
	23·0								1·770		
	23·5								2·160		
	24·0								2·560		
	24·5								3·080	1·520	
	25·0									1·700	
	25·5									1·910	
	26·0									2·170	
	26·5									2·500	
	27·0									2·940	
	27·5										
	28·0										1·840
	28·5										2·050
	29·0										2·380
	29·5										2·800
	30·0										3·320

General tree assortment table for conifers
Volumes (OB) to various top diameters underbark

TABLE 4

Under-bark top dia (in cm)	Length in metres	Height class in metres (height to cm in brackets)									
		9 (6·5)	12 (9·5)	15 (12)	18 (15)	21 (18)	24 (21)	27 (24)	30 (27)	33 (30)	36 (33)
18	2·0	0·075	0·065								
-	2·5	0·095	0·085	0·080							
	3·0	0·120	0·105	0·095							
	3·5	0·145	0·130	0·115	0·110						
	4·0	0·170	0·155	0·140	0·125	0·120					
	4·5		0·190	0·165	0·145	0·140	0·135				
	5·0		0·220	0·190	0·165	0·160	0·155				
	5·5		0·255	0·220	0·190	0·180	0·175				
	6·0		0·295	0·250	0·215	0·200	0·195				
	6·5		0·340	0·290	0·245	0·225	0·215				
	7·0			0·330	0·275	0·250	0·235				
	7·5			0·370	0·305	0·275	0·260				
	8·0			0·420	0·340	0·305	0·285				
	8·5			0·480	0·380	0·335	0·310				
	9·0			0·550	0·430	0·370	0·335				
	9·5			0·625	0·475	0·405	0·360				
	10·0			0·720	0·530	0·440	0·390				
	10·5				0·600	0·485	0·420				
	11·0				0·690	0·535	0·455				
	11·5				0·820	0·585	0·490				
	12·0					0·650	0·525				
	12·5					0·720	0·565				
	13·0					0·795	0·610				
	13·5					0·890	0·660				
	14·0					1·010	0·720				
	14·5					1·195	0·785				
	15·0					1·525	0·860	0·720			
	15·5						0·960	0·770			
	16·0						1·070	0·830			
	16·5						1·200	0·900			
	17·0						1·410	0·980			
	17·5						1·730	1·070	0·830		
	18·0							1·180	0·890		
	18·5							1·310	0·960		
	19·0							1·470	1·040		
	19·5							1·650	1·130		
	20·0							1·990	1·220		
	20·5							2·350	1·320		
	21·0								1·450		
	21·5								1·600		
	22·0								1·800		
	22·5								2·090		
	23·0								2·500	1·580	
	23·5								2·890	1·690	
	24·0									1·810	
	24·5									1·950	
	25·0									2·160	
	25·5									2·430	
	26·0									2·880	1·770
	26·5									3·430	1·890
	27·0										2·060
	27·5										2·260
	28·0										2·530
	28·5										2·940
	29·0										3·420

Standard Times for Chainsaw Crosscutting Pole Lengths at Roadside into Scottish Pulpwood and/or STP Billets and Sawlogs
East Scotland

1 Conditions

The time-studies on which these standard times are based were carried out in East Scotland. Providing that the conditions, job specification, tools and equipment and the working method are as shown, they will apply in other areas.

The standard times apply to crosscutting carried out under the following conditions:

a. Crosscutting conifer species at roadside by one man (see Appendix for working method).

b. Poles stacked, in heaps of approximately 1·5 m³ volume, on bearers and parallel to the road.

c. Poles in an acceptable condition to meet the product specification, ie root spurs removed and clean snedded with only occasional stubs or branches to be removed during crosscutting.

d. Produce to be carried not more than an average of 3 m (see paragraph 7b where the distance carried is up to 7 m).

e. Pole volume to be determined by the modified tariff system described in Forestry Commission Booklet No 39, Procedure 10.

f. Poles to be reasonably clean and free from excessive dirt (see also paragraph 7c).

2 Job specification

The standard times are for the following work:

a. Crosscutting of product mixes which are combinations of the following:
  (i) Sawlogs—random or specified lengths with a minimum top diameter of 13 cm;
  (ii) Scottish pulpwood—standard length of 3 m with downward tolerance to 2·4 m, over bark diameter range 5 cm to 41 cm;
  (iii) STP (Scottish Timber Products), category 1 (spruce, pine or fir)—length 1·8 m to 2·0 m with downward tolerance to 1·5 m, over bark diameter range 5 cm to 35 cm;

273

(iv) STP, category 2 (larch)—length 1·8 m to 2·0 m or 2·5 m to 3·0 m, over bark diameter range 5 cm to 35 cm.

b. All saw cuts to be made squarely across the pole.

c. All produce (except large sawlogs) to be suitably stacked for mechanical loading, parallel to the road if space is available and at right angles if it is not. Average carrying distance should not exceed 3 m per piece (see paragraph 7b if it is not possible to achieve this) and stack height should not exceed 1·3 m.

d. Branches and stubs remaining after snedding to be removed.

e. Roads and roadside drains to be left clear of all brash and waste.

f. Should the crosscutter be required to assist with unchokering, extra time will be required (see paragraph 7a).

3 **Tools and equipment** (including safety equipment)
   a. Lightweight chainsaw of an approved type together with maintenance tools and fuel cans.

   b. Logger's belt and spring-loaded tape.

   c. Measuring rod.

   d. Pulphook and timber tongs of an approved type.

   e. Pitchfork (or similar tool) for clearing brash.

   f. Protective clothing as specified by the safety officer, viz:
   Safety helmet (complying with BS 2826, 2095 or 5240).
   Ear protectors.
   Mesh visor.
   Gloves, incorporating ballistic nylon guard on back of left hand.
   Standard pattern Forestry Commission chainsaw operator's trousers.
   Nylon leg guard (left leg only).
   Safety boots, incorporating inner ballistic nylon guard.

4 **Allowances**
   a. For contingencies and work other than that performed on individual poles, eg refuelling, day-to-day maintenance of the saw, clearing brash from roads etc, 20% of the time spent measuring, crosscutting and stacking.

   N.B. Mechanical breakdowns in excess of 15 minutes are not included in the allowances.

   b. For personal needs and rest 19–30% of the total working time.

5 **Method of using the standard time table**
   a. Obtain the mean pole volume using the tariff system.

   b. Obtain the mean number of pieces cut per pole. Normally a sample of 10 loads will give a sufficiently accurate result for any

274

area being worked provided that the crop is uniform and that product mix remains the same.

c. Select the appropriate standard time from paragraph 6 and then modify as necessary with reference to paragraph 7.

Example 1
Crosscutting into sawlogs, Scottish pulpwood and STP
Mean pole volume 0·08 m³
Mean number of pieces cut per pole 2·4
Conditions, job specification and tools and equipment as shown in paragraphs 1, 2 and 3 of this standard time table.
Time per tree (from paragraph 6) = 1·51 SMs

Example 2
Crosscutting into sawlogs, Scottish pulpwood and STP
Mean pole volume 0·08 m³
Mean number of pieces cut per pole 2·5
Conditions, job specification and tools and equipment as shown in paragraphs 1, 2 and 3 of this standard time table
Time per tree (from paragraph 6)

$$\text{Interpolated } 1·51 + \frac{(1·59 - 1·51)}{2} = 1·55 \text{ SMs}$$

Example 3
Crosscutting into sawlogs, Scottish pulpwood and STP
Mean pole volume 0·08 m³
Mean number of pieces cut per pole 2·4
Average carrying distance 6 m per piece
Crosscutter assists with unchokering
Other conditions, job specification and tools and equipment as shown in paragraphs 1, 2 and 3 of this standard time table
Time per tree (from paragraph 6)   = 1·51 SMs
Extra time for excessive carrying
(from paragraph 7b)
The standard times include carrying up to 3 m
∴ 6 m − 3 m = 3 m × 0·10 SMs per pole   = 0·30 SMs
Extra time for unchokering
(from paragraph 7a)   = 0·38 SMs
                              Total = 2·19 SMs per pole

Example 4
Crosscutting into sawlogs, Scottish pulpwood and STP
Mean pole volume 0·08 m³
Mean number of pieces cut per pole 2·4
Crosscutter assists with unchokering
Very dirty poles
Other conditions, job specification and tools and equipment as shown in paragraphs 1, 2 and 3 of this standard time table.

Time per pole (from paragraph 6)   = 1·51 SMs
Extra time for dirty poles
(from paragraph 7c)

$$\frac{1·51}{100} \times 5$$ = 0·08 SMs

Extra time for unchokering
(from paragraph 7a)   = 0·38 SMs

Total = 1·97 SMs per pole

N.B. This standard time table can be used to predict the average standard time per piece for a given situation. The standard time per pole is assessed in the way shown above and is then simply divided by the mean number of pieces per pole.

eg Standard time per pole   = 1·97 SMs
    Mean number of pieces per pole = 2·4
    Standard time per piece   = $\dfrac{1·97}{2·4}$

    Standard time per piece   = 0·82 SMs

It should be noted that this standard time is the average time per piece and that no single piece is likely to take exactly that amount of time. However, providing that the time per pole and the average number of pieces per pole have been correctly assessed, it is an accurate basis for payment purposes; the standard time per pole remains the same with either method.

6 Standard time for crosscutting at roadside

Mean pole volume in m^3	Standard minutes per pole							
	Mean number of pieces cut per pole							
	1·8	2·0	2·2	2·4	2·6	2·8	3·0	3·2
0·04	0·89	0·98	1·05	1·11	1·20	1·27	1·34	1·42
0·05	0·90	0·99	1·06	1·13	1·21	1·28	1·35	1·43
0·06	0·97	1·04	1·12	1·19	1·28	1·35	1·42	1·50
0·07		1·18	1·25	1·33	1·40	1·47	1·56	1·63
0·08		1·37	1·44	1·51	1·59	1·66	1·73	1·81
0·09		1·57	1·63	1·71	1·78	1·86	1·93	2·00
0·10		1·77	1·83	1·89	1·98	2·04	2·12	2·18
0·11		1·91	1·98	2·04	2·12	2·19	2·26	2·33
0·12		2·00	2·06	2·12	2·20	2·26	2·33	2·41

N.B. Interpolate as necessary.

7 Modifications and variations to the standard times

N.B. These modifications and variations are only to be applied when the prevailing conditions, or job specification, differ from those listed in paragraphs 1 and 2.

a. *Unchokering carried out by crosscutter*
When the crosscutter assists the skidder operator with unchokering by removing detachable chokers.

Add 0·38 standard minutes per pole to the times shown in paragraph 6.

b. *Extra carrying distance*
The average carrying distance assumed in this standard time table is 3 m. For every metre produce is carried in excess of this

Add 0·10 standard minutes per pole to the times shown in paragraph 6.

N.B. Average carrying distance should, under no circumstances, exceed 7 m. For the method of making this allowance see example 3, paragraph 5.

c. *Dirty poles*
When poles are excessively gritty, resulting in extra sharpening of the chain

Add 5% to the times shown in paragraph 6.

N.B. This modification must be made before either of the additions shown above are made. For the method of making this allowance see example 4, paragraph 5.

Working method—conversion at roadside

Unchokering

1. By tractor driver alone—method irrelevant to crosscutter.
2. By tractor driver plus crosscutter.

   a. *No chainsaw used*
      - (i) Tractor driver drops load and drives on at least 2 m;
      - (ii) Crosscutter removes terminal pins and stands clear;
      - (iii) Tractor driver winches in, freeing winch ropes from all chokers;
      - (iv) Both remove and stow chokers.

   b. *Chainsaw used*
      - (i) Tractor driver drops load and drives on at least 2 m;
      - (ii) Crosscutter saws through all poles immediately below the chokers;
      - (iii) Crosscutter withdraws terminal pins;
      - (iv) Tractor driver winches in, freeing winch ropes from all chokers;
      - (v) Both remove and stow chokers.

Crosscutting

1. Walk to heap of poles.
2. Measure first sawlog (if none, step 5 applies).
3. Crosscut first sawlog.
4. Repeat for all sawlogs within easy reach of one side of the heap without moving produce or placing both feet on produce.
5. Measure first piece of pulp.
6. Crosscut first piece of pulp.
7. Repeat for all pulp within easy reach of one side of the heap without moving produce or placing both feet on produce.
8. Turn and sned pieces if required.
9. Unless stacking of converted produce is only possible on one side of the heap, repeat steps 2–8 on the other side of the heap.
10. Stack all converted produce and move aside any waste which may impede further crosscutting.

    Repeat steps 1–10 until the heap is finished.

    N.B. See notes on method following.

Notes on method

Unchokering

It is assumed that in first thinnings tip first extraction will be practised. If for any reason poles are extracted butt first then method a. may be used. Method b. will usually be the most efficient.

The percentage of waste entailed will be minimal and in most cases would have been waste anyway or would have needed snedding to remove the whorl of branch stubs left on for chokering. If random length sawlogs are being produced, the crosscutter should, if possible, cut them off before unchokering and well clear of the pulp conversion site. The remainder of the load, containing the pulp element, should then be dropped on the grid.

Crosscutting

Measurement of short sawlogs will normally be by measuring rod. Measurement of pulp, both STP and SPPM, may be by eye or by measuring rod. Some men have the ability to judge by eye, others prefer to reduce the level of concentration required by using a measuring rod.

Measurement and crosscutting of pulp should start at the butt end of the heap and progress to the tip end. The butt pieces of pulp should be cut from all the easily reached poles followed by the second pieces of pulp from the same poles etc. In this way conversion of these poles will be completed in one pass rather than working along an individual pole to the tip then walking back to the butt of the next. If it is intended to maximise Scottish pulp after removing the sawlogs, the crosscutter should ignore the possibility of producing STP until the entire heap has been converted. Any waste remaining with a length greater than 1·5m (1·8m in the case of larch) should be stacked with the butt ends flush. If this stack is then crosscut to a maximum length of 2m it is then all within the STP specification.

It is essential that the crosscutter should keep at least one foot on firm, level ground whilst using the chainsaw. It is specified in the Output Guide that the heaps of poles should be parallel to the road. To stack the converted produce parallel to the road requires the least movement of produce and thus the least energy output from the crosscutter. It is recognised that parallel stacking has the disadvantage of using more space for a given volume of produce, but if the space is available it should be used to the crosscutter's advantage.

Chainsaw Crosscutting at Roadside Into 3 m Pulpwood Lengths With Millwood, Chipper Logs and Stake Lengths (Measuring With a Rod and Scribe)
South Wales

1 **Conditions**

The Output Guide applies to the following conditions:

  a. Crosscutting conifer species at roadside by one man (see Appendix I for method).

  b. Poles stacked parallel to the road on bearers.

  c. Poles in an acceptable condition to meet the product specification, ie root spurs taken off and clean snedded with only occasional stubs or branches to be removed during crosscutting (but see para 7b).

2 **Job specification**

The Output Guide applies to the following work:

  a. Product mixes are combinations of the following:

    (i) Millwood—short lengths 1·83 m–3·05 m with a minimum top diameter of 16 cm under bark.

    (ii) Chipper logs—lengths 1·83 m–3·07 m with a top diameter under bark of 13 cm–18 cm.

    (iii) Pulpwood billets—length 3·05 m (some down to 2·44 m) with a minimum top diameter of 8 cm over bark and a maximum butt diameter of 30 cm over bark (up to 5% butt diameters up to 38 cm over bark).

    (iv) Stake lengths—1·65 m.

  b. All saw cuts to be made squarely across the pole.

  c. All produce except millwood to be suitably stacked for mechanical loading, parallel to the road (if millwood needs to be moved see para 7d).

  d. Roads and drains to be left clear of all brash and waste.

3 **Tools and equipment** (including safety equipment)

  a. Lightweight anti-vibration chainsaw with maintenance tools and fuel cans (see para 7e).

b. Measuring rod (3 m in length, graduated), scribe and rule for measuring top diameters.

c. Felt markers and crayons (see para 7c).

d. One pulphook and one timber tongs.

e. Breaking bar with cant-hook attachment (for moving millwood during measure and mark).

f. Protective clothing viz:
Safety helmet BS 2826, 2095 or 5240.
Eye protection.
Ear defenders.
Gloves, incorporating ballistic nylon lining on back of left hand.
Nylon leg guard.
Safety boots, incorporating inner ballistic nylon guard.

4 Allowances

Included in the times

a. Contingencies and other work:
   (i) For contingencies and work other than walk and prepare, measure and crosscut, eg tool maintenance, moving poles, walking between stacks etc, 54% of the time spent on walk and prepare, measure and crosscut.
   (ii) For contingencies and work other than stacking, eg tidy stacks, walk between stacks etc, 14% of the time spent on stacking.

b. Personal needs and rest:
   (i) For walk and prepare, measure and crosscut, 23% of the total working time for these operations.
   (ii) For stacking, from 24% to 64% of the total stacking time according to the volumes of the pieces stacked.

5 Method of using the Output Guide

Use table 6A to obtain the time for walk and prepare, measure and crosscut per pole. To obtain the time, the mean pole volume and the mean number of pieces cut per pole are required.

The mean pole volume will normally be the tariff volume (see FC Booklet No 36) unless primary conversion of millwood has been done at stump. For guidance in determining millwood content see Appendix III.

The mean number of pieces cut per pole can only be calculated accurately at the end of the job by dividing total number of pieces by total number of poles. Alternatively samples can be taken during working but for accuracy they need to be spread through the crop.

It is suggested that, unless the supervisor's experience suggests otherwise the mean figure in column 3 may be taken as a starting point and adjustment can be made to the times as information on the number of pieces is gathered.

Note: If the mean number of pieces per pole differs from that in column 3 of the table *add* or *subtract* 0·019 standard minutes for each tenth of a piece different from the tabulated mean number of pieces.

Use table 6B to obtain the time for stacking. To obtain the time for stacking per pole, mean number of pieces stacked per pole, and mean volume of the pieces stacked are required.

Where no millwood is produced—this simply means dividing the original pole volume by the number of pieces cut per pole. The time for stacking appropriate to this mean volume per piece is then selected from table 6B; the time per pole is obtained by *multiplying* this figure by the mean number of pieces per pole (see example 1 in Appendix II).

Where millwood is produced—the mean pole volume must be reduced by the total volume of millwood cut per pole before table 6B is used and the number of pieces must be reduced by the number of millwood pieces cut per pole (see example 2 in Appendix II).

The time per pole for the whole operation is obtained by the addition of the time for walk and prepare, measure, crosscut (table 6A) and times for stacking (table 6B). For worked examples see Appendix II.

6A Time in standard minutes per pole for walk and prepare, measure and crosscut

Volume of pole in m³	Guidance time in SMs	Mean number of pieces per pole
0·08	1·38	2·9
0·10	1·55	3·1
0·14	1·85	3·5
0·18	2·09	3·9
0·22	2·32	4·2
0·26	2·52	4·5
0·30	2·72	4·8
0·34	2·88	5·0
0·38	3·05	5·1
0·42	3·21	5·3
0·46	3·36	5·5
0·50	3·51	5·6
0·54	3·64	5·7

Volume of pole in m³	Guidance time in SMs	Mean number of pieces per pole
0·58	3·79	5·8
0·62	3·90	5·9
0·66	4·04	6·0
0·70	4·17	6·0
0·74	4·28	6·1
0·78	4·39	6·1
0·82	4·51	6·1
0·86	4·62	6·2

Add or **subtract** 0·019 standard minutes for each tenth of a piece more or less than the mean number of pieces shown in column 3.

6B   Time in standard minutes per mean piece for stacking

Volume of mean piece in m³	Guidance time in SMs
0·015	0·22
0·020	0·27
0·030	0·35
0·040	0·42
0·050	0·47
0·060	0·52
0·070	0·57

Note:

Where crosscutting is carried out in conjunction with extraction and millwood dimensions are such that they need to be moved by log rolling blade, delays would sometimes be incurred at the crosscutting stage.

The delays must be assessed separately because they depend on the extraction distance and no allowance has been included in the Output Guide.

7   Modifications and variations to the Output Guide

N.B. These modifications and variations are only to be applied when the prevailing conditions or job specification differ from those listed in paragraphs 1 and 2.

a. *Dirty poles*

When poles are excessively gritty resulting in extra sharpening of the chain

Add 5% to the total time per pole.

b. *Extra snedding*

Where extra snedding is necessary during crosscutting

Add 10% to the total time per pole.

(Extra snedding is usually required where large poles are deliberately left unturned at the felling and snedding stage.)

c. *Measure, mark and book millwood pieces*

For measuring top diameter (UB) marking these top diameters on the log with a crayon or felt marker and entering the information in a book or on a form as required

Add 0·50 standard minutes per piece.

d. *Minimum movement of millwood pieces*

If it is necessary to move millwood pieces by hand so that traffic is not impeded

Add 0·20 standard minutes per millwood piece moved (when movement of millwood is by rolling it to the side of the road).

e. *Use of saws of less than 68 cc*

If a lower powered saw such as a Jonsered 621 (56 cc) or Husqvarna 263 (63 cc) is used for crosscutting, make the following additions to the times:

When using table 6A

Add 5% if the mean pole is between 0·40 m³ and 0·80 m³,

Add 10% if the mean pole is greater than 0·80 m³.

Recommended working method (one-man working)

1. Pick up measuring rod and scribe.
2. Walk to stack.
3. Measure from butt end and scribe as many pieces as possible in the stack.
4. Aside measuring rod and scribe.
5. Pick up saw.
6. Walk to stack.
7. Crosscut scribed pieces.
8. Aside saw.
9. Stack crosscut pieces.

Repeat sequence until stack is completed.

Note:
(i)   Stacking of millwood pieces is not included in the Output Guide.
(ii) Pieces in excess of $0\cdot050\,m^3$ in volume should be rolled or lifted one end at a time on to the stack.

Method of using the Output Guide

Example 1

Average pole volume from tariff $= 0\cdot18\,m^3$

Mean number of pieces cut per pole $= 3\cdot7$ (no millwood)

Time for walk and prepare, measure and crosscut per pole

$=$ Time from table 6A for $0\cdot18\,m^3$ pole-adjustment for number of pieces cut

$= 2\cdot09$ standard minutes $-[0\cdot019 \times (3\cdot9-3\cdot7)]$ standard minutes

$= 2\cdot09$ standard minutes $-0\cdot019 \times 2$ standard minutes

$= 2\cdot09 - 0\cdot038$

$= 2\cdot05$ standard minutes

Standard time for stacking (for a pole which has no millwood produced)

Mean volume per piece stacked

$$= \frac{\text{average pole volume from tariff}}{\text{mean number of pieces cut}}$$

$$= \frac{0\cdot18\,m^3}{3\cdot7}$$

$= 0\cdot049\,m^3$

Time for stacking per piece from table 6B $= 0\cdot47$ standard minutes

Time for stacking per pole $= 0.47 \times 3.7$ standard minutes
$= 1.74$ standard minutes
Total time per pole = time for measure and crosscut plus time for stacking
$= 2.05 + 1.74$
Total time per pole $= 3.79$ standard minutes

Example 2
Average pole volume from tariff $= 0.340\,\text{m}^3$
Mean number of pieces cut per pole $= 4.9$ pieces
Total volume of millwood per pole $= 0.167\,\text{m}^3$
Total number of pieces of millwood per pole $= 1.7$ pieces

Time for measure and crosscut
= Time from table 6A for $0.340\,\text{m}^3$ pole-adjustments for number of pieces cut
$= 2.88 - (0.019 \times 1)$
$= 2.88 - 0.019$ standard minutes
$= 2.86$ standard minutes

Time for stacking
Mean volume per piece stacked

$$= \frac{\text{mean pole volume minus millwood volume}}{\text{assessed number of pieces minus millwood pieces per pole}}$$

$$= \frac{0.340 - 0.167}{4.9 - 1.7}$$

$$= \frac{0.173}{3.2}$$

$= 0.054\,\text{m}^3$
Time for stacking per piece from table 6B
$= 0.49$ standard minutes
Time for stacking per pole
$= 0.49 \times 3.2$ standard minutes
$= 1.57$ standard minutes
Total time per pole
= time for measure and crosscut plus time for stacking
$= 2.86 + 1.57$ standard minutes
$= 4.43$ standard minutes

APPENDIX III

A method of assessing timber volume and number of pieces of millwood per tree

1. Tables 1, 2 and 3 are tree assortment tables giving length to top diameter under bark of 16, 18 and 24 cm respectively for trees of various breast height diameter classes and by height classes

(height class 18 means trees with total heights between $16\frac{1}{2}$ and $19\frac{1}{2}$ metres etc).

2. Tables 4, 5 and 6 show the volume in cubic metres by millwood length and tree height class for top diameters under bark of 16, 18 and 24 cm respectively.

Notes:
(i)   Tables 1–6. The top diameter under bark is the diameter to which millwood is cut.
(ii) Tables 1–6. The height class refers to the overall height of the tree. The most accurate method of finding the height of each size of tree is by making comparison with the sample trees on the U15(D) using lengths to 7 cm which are shown in brackets under the height classes on the tables. Alternatively, assess the top height of the crop and drop a class to obtain a figure for the mean tree to be felled which is quicker but less accurate.
(iii) Tables 1, 2 and 3. The breast height diameter class is taken from the girthing record (C) on the U15 (tariff sheet).

3. *Method of working*
(i)   For each size of tree on the girthing record on the U15(C) read the length to the required under bark top diameter from the appropriate table (1, 2 or 3) against the appropriate breast height diameter and tree height class.
(ii) From the length so obtained assess the number and lengths of pieces of millwood that can be cut from each size of tree and add their lengths together to obtain the length of timber to be obtained from the trees, which may be the same as, or less than the length obtained from the table.
(iii) Using the length of millwood in each size of tree obtain its volume from table 4, 5 or 6 as appropriate, using the same tree height class (or height to 7 cm) as in para 1.
(iv) For each breast height diameter size of tree multiply the number of pieces of timber and the volume of timber in the tree by the number of trees of that size in the stand. Sum the total number of pieces and the total volume of timber in all the trees to be felled and divide each total by the number of trees to obtain the mean number of pieces and the mean volume of timber per tree.

Example
Millwood specifications
1. 1·85 metres in length with a top diameter of 16 cm under bark or more.
2. 2·65 metres in length with a top diameter of 16 cm under bark or more.

288

Breast height dia cm	Number of trees	Height class in metres	Lengths to 16 cm under bark in metres	Estimated number of pieces per pole	Total length of estimated pieces in m	Vol of estimated pieces in pole (m³)	Total number of pieces in BH dia class (2×5)	Total volume in BH dia class (m³) (2×7)
(1)	(2)	(3)	(4)	(5)	(6)	(7)	(8)	(9)
20	5	18	5·3	2(2 × 2·65)	5·3	0·149	10	0·745
21	—	—	—	—	—	—	—	—
22	7	18	7·1	3(1 × 2·65) (2 × 1·85)	6·4	0·182	21	1·274
23	8	18	7·9	3(2 × 2·65) (1 × 1·85)	7·2	0·215	24	1·720
24	10	18	8·5	4(1 × 2·65) (3 × 1·85)	8·2	0·272	40	2·720
25	14	18	9·1	4(2 × 2·65) (2 × 1·85)	9·0	0·325	56	4·550
26	20	18	9·6	4(2 × 2·65) (2 × 1·85)	9·0	0·325	80	6·500
27	22	18	10·0	4(3 × 2·65) (1 × 1·85)	9·8	0·392	88	8·624
28	15	18	10·4	4(3 × 2·65) (1 × 1·85)	9·8	0·392	60	5·880
29	7	18	10·7	4(4 × 2·65)	10·6	0·472	28	3·304
30	6	18	11·0	4(4 × 2·65)	10·6	0·472	24	2·832
31	5	18	11·3	5(2 × 2·65) (3 × 1·85)	10·9	0·508	25	2·540
32	1	18	11·5	5(2 × 2·65) (3 × 1·85)	10·9	0·508	5	0·508
Total	120						461	41·197

Total volume of millwood per tree $= \dfrac{\text{Total col 9}}{\text{Total col 2}} = \dfrac{41\cdot197}{120} = 0\cdot343\,\text{m}^3$

Number of pieces of millwood per tree $= \dfrac{\text{Total col 8}}{\text{Total col 2}} = \dfrac{461}{120} = 3\cdot84$

Notes:

Column 1   Taken directly from Girthing Record (U15(C)).
Column 2   Taken directly from Girthing Record (U15(C)).
Column 4   From table 1 (breast height diameter and appropriate height class).
Column 5   Probable utilisation of the length according to the millwood specifications (figures in brackets are an explanation of the number of pieces).
Column 6   The total length of the pieces in column 5.
Column 7   From table 4 (the appropriate volume for the length quoted in column 6 according to height class).
Column 8   Column 5 multiplied by column 2.
Column 9   Column 7 multiplied by column 2.

Table 1 General tree assortment table for conifers
Length to 16 cm top diameter under bark

Breast height dia class (in cm)	Height class in metres (height to 7 cm in brackets)									
	9 (6·5)	12 (9·5)	15 (12)	18 (15)	21 (18)	24 (21)	27 (24)	30 (27)	33 (30)	36 (33)
20	2·7	3·3	4·5	5·3	6·3	8·0				
21	3·1	3·9	5·1	6·3	7·3	9·2				
22	3·5	4·4	5·7	7·1	8·3	10·3				
23	3·8	4·9	6·3	7·9	9·2	11·4				
24	4·3	5·3	6·8	8·5	10·0	12·3				
25	4·6	5·8	7·3	9·1	10·7	13·0				
26		6·2	7·7	9·6	11·3	13·7				
27		6·5	8·0	10·0	11·8	14·2				
28		6·8	8·4	10·4	12·2	14·7				
29		7·1	8·7	10·7	12·6	15·1				
30		7·3	9·0	11·0	13·0	15·5	17·5	20·0		
31			9·2	11·3	13·3	15·8	17·8	20·3		
32			9·4	11·5	13·6	16·1	18·1	20·6		
33			9·6	11·7	13·9	16·4	18·4	20·9		
34			9·8	11·9	14·2	16·6	18·7	21·2		
35			10·0	12·1	14·4	16·8	18·9	21·4		
36			10·2	12·3	14·6	17·0	19·1	21·6		
37			10·4	12·4	14·8	17·2	19·3	21·8		
38			10·5	12·6	14·9	17·3	19·5	22·0		
39			10·6	12·7	15·0	17·4	19·7	22·2		
40			10·8	12·8	15·1	17·5	19·8	22·4	24·8	28·1
41					15·2	17·6	20·0	22·5	25·0	28·3
42					15·2	17·7	20·1	22·7	25·2	28·5
43					15·3	17·8	20·2	22·8	25·4	28·6
44					15·4	17·9	20·3	23·0	25·6	28·7
45					15·4	18·0	20·4	23·1	25·8	28·8
46					15·5	18·0	20·5	23·2	25·9	29·0
47					15·5	18·1	20·6	23·3	26·0	29·1
48					15·6	18·1	20·7	23·4	26·2	29·2
49					15·6	18·2	20·8	23·5	26·3	29·3
50					15·7	18·3	20·8	23·6	26·4	29·4
51						18·3	20·9	23·7	26·5	29·5
52						18·3	21·0	23·7	26·6	29·6
53						18·4	21·0	23·8	26·7	29·7
54						18·4	21·1	23·9	26·8	29·7
55						18·5	21·1	24·0	26·9	29·8
56							21·2	24·1	27·0	29·9
57							21·2	24·1	27·0	30·0
58							21·3	24·2	27·1	30·0
59							21·4	24·3	27·2	30·1
60							21·5	24·4	27·2	30·2

Table 2 General tree assortment table for conifers
Length to 18 cm top diameter under bark

Breast height dia class (in cm)	Height class in metres (height to 7 cm in brackets)									
	9 (6·5)	12 (9·5)	15 (12)	18 (15)	21 (18)	24 (21)	27 (24)	30 (27)	33 (30)	36 (33)
20	2·00	2·10	2·60	3·00	3·20	4·00	3·90			
21	2·40	2·75	3·50	4·20	4·90	5·50	6·05			
22	2·80	3·30	4·20	5·20	6·10	6·90	7·95			
23	3·25	3·85	4·90	6·00	7·20	8·20	9·50			
24	3·60	4·35	5·50	6·75	8·05	9·45	10·95			
25	4·00	4·80	6·10	7·40	8·85	10·40	12·20			
26		5·20	6·50	7·95	9·45	11·45	13·00			
27		5·60	6·95	8·45	10·00	12·30	13·70			
28		6·00	7·30	8·90	10·50	13·00	14·40			
29		6·40	7·70	9·30	11·00	13·50	15·00			
30		6·70	8·00	9·70	11·40	14·00	15·50	17·90		
31			8·30	10·00	11·80	14·40	16·05	18·40		
32			8·60	10·35	12·10	14·70	16·45	18·80		
33			8·90	10·65	12·40	15·00	16·85	19·20		
34			9·10	10·90	12·75	15·30	17·20	19·60		
35			9·30	11·20	13·00	15·50	17·50	19·90		
36			9·55	11·40	13·25	15·75	17·80	20·25		
37			9·70	11·60	13·50	16·00	18·00	20·50		
38			9·90	11·75	13·70	16·15	18·25	20·80		
39			10·10	11·90	13·90	16·35	18·50	21·00		
40			10·20	12·10	14·10	16·50	18·70	21·20	23·60	26·80
41					14·25	16·65	18·90	21·45	23·80	26·95
42					14·40	16·80	19·10	21·65	24·05	27·15
43					14·50	16·90	19·25	21·80	24·30	27·30
44					14·60	17·05	19·40	22·00	24·50	27·50
45					14·70	17·10	19·50	22·20	24·70	27·60
46					14·80	17·25	19·65	22·35	24·90	27·75
47					14·90	17·35	19·80	22·45	25·10	27·90
48					14·95	17·45	19·90	22·60	25·30	28·10
49					15·00	17·55	20·00	22·70	25·40	28·25
50					15·00	17·60	20·10	22·80	25·60	28·40
51						17·70	20·20	22·95	25·70	28·50
52						17·75	20·25	23·05	25·85	28·65
53						17·80	20·35	23·15	25·95	28·75
54						17·90	20·45	23·25	26·10	28·85
55						17·90	20·50	23·30	26·20	29·00
56								23·40	26·30	29·10
57								23·50	26·40	29·20
58								23·60	26·50	29·30
59								23·65	26·55	29·40
60								23·70	26·60	29·50

Table 3 General tree assortment table for conifers
Length to 24 cm top diameter under bark

Breast height dia class (in cm)	Height class in metres (height to 7 cm in brackets)								
	12 (9·5)	15 (12)	18 (15)	21 (18)	24 (21)	27 (24)	30 (27)	33 (30)	36 (33)
25	1·30	1·30	1·30	1·30	1·30	1·30	1·30		
26	1·80	2·00	2·20	2·50	2·70	2·60	3·70		
27	2·20	2·60	2·90	3·40	3·90	3·90	5·50		
28	2·60	3·20	3·70	4·30	4·90	5·10	6·90		
29	3·00	3·80	4·40	5·10	5·90	6·20	8·00		
30	3·40	4·30	5·00	5·80	6·70	7·30	9·10		
31		4·80	5·60	6·50	7·40	8·15	9·90		
32		5·20	6·10	7·05	8·15	9·00	10·70		
33		5·60	6·70	7·60	8·80	9·85	11·50		
34		5·95	7·10	8·10	9·40	10·60	12·25		
35		6·30	7·60	8·50	9·90	11·10	12·80	15·30	17·10
36		6·65	8·00	9·00	10·50	11·80	13·60	16·00	17·90
37		6·90	8·40	9·35	10·95	12·40	14·30	16·60	18·60
38		7·20	8·70	9·75	11·40	12·90	14·90	17·20	19·20
39		7·45	9·00	10·10	11·80	13·40	15·45	17·70	19·80
40		7·70	9·40	10·40	12·20	13·90	15·90	18·10	20·30
41				10·75	12·50	14·25	16·40	18·60	20·80
42				11·05	12·85	14·60	16·80	19·00	21·30
43				11·30	13·15	15·00	17·25	19·40	21·70
44				11·55	13·40	15·30	17·60	19·80	22·15
45				11·80	13·65	15·60	18·00	20·20	22·60
46				11·95	13·90	15·90	18·30	20·50	23·00
47				12·10	14·10	16·20	18·60	20·80	23·40
48				12·25	14·30	16·50	18·85	21·10	23·70
49				12·35	14·45	16·75	19·15	21·40	24·00
50				12·40	14·60	17·00	19·30	21·80	24·30
51					14·80	17·20	19·55	22·00	24·70
52					15·00	17·40	19·80	22·25	25·00
53					15·20	17·55	19·95	22·50	25·25
54					15·35	17·70	20·15	22·75	25·50
55					15·50	17·80	20·30	22·95	25·70
56						18·00	20·40	23·20	26·00
57						18·15	20·60	23·35	26·20
58						18·30	20·75	23·50	26·40
59						18·40	20·85	23·65	26·55
60						18·50	21·00	23·80	26·70

Table 4 General tree assortment table for conifers
Volume (OB) to 16 cm top diameter under bark

Length in metres	Height class in metres (height to 7 cm in brackets)									
	9 (6·5)	12 (9·5)	15 (12)	18 (15)	21 (18)	24 (21)	27 (24)	30 (27)	33 (30)	36 (33)
2·0	0·055									
2·5	0·080									
3·0	0·100	0·085								
3·5	0·120	0·105								
4·0	0·150	0·125								
4·5	0·180	0·150	0·130							
5·0	.	0·180	0·145	0·140						
5·5		0·205	0·170	0·155						
6·0		0·240	0·195	0·170	0·165					
6·5		0·280	0·220	0·185	0·180					
7·0		0·320	0·250	0·205	0·200					
7·5			0·285	0·230	0·220					
8·0			0·330	0·260	0·240	0·225				
8·5			0·380	0·290	0·260	0·235				
9·0			0·430	0·325	0·280	0·250				
9·5			0·500	0·365	0·305	0·270				
10·0			0·580	0·410	0·335	0·290				
10·5			0·690	0·460	0·365	0·310				
11·0				0·520	0·400	0·330				
11·5				0·590	0·440	0·355				
12·0				0·680	0·485	0·385				
12·5				0·790	0·535	0·415				
13·0					0·595	0·450				
13·5					0·660	0·490				
14·0					0·740	0·535				
14·5					0·835	0·585				
15·0					0·990	0·640				
15·5					1·280	0·710				
16·0						0·785				
16·5						0·875				
17·0						1·000				
17·5						1·180	0·790			
18·0						1·460	0·850			
18·5						2·000	0·940			
19·0							1·040			
19·5							1·190			
20·0							1·390	0·905		
20·5							1·700	0·980		
21·0							2·120	1·060		
21·5							2·690	1·175		
22·0								1·310		
22·5								1·500		
23·0								1·770		
23·5								2·160		
24·0								2·560		
24·5								3·080	1·520	
25·0									1·700	
25·5									1·910	
26·0									2·170	
26·5									2·500	
27·0									2·940	
27·5										
28·0										1·840
28·5										2·050
29·0										2·380
29·5										2·800
30·0										3·320

293

Table 5 General tree assortment table for conifers
Volume (OB) to 18 cm top diameter under bark

Length in metres	Height class in metres (height to 7 cm in brackets)									
	9 (6·5)	12 (9·5)	15 (12)	18 (15)	21 (18)	24 (21)	27 (24)	30 (27)	33 (30)	36 (33)
2·0	0·065	0·069								
2·5	0·089	0·085	0·079							
3·0	0·115	0·105	0·095	0·093	0·092					
3·5	0·143	0·129	0·117	0·112	0·110					
4·0	0·172	0·157	0·138	0·130	0·128	0·122	0·122			
4·5		0·185	0·160	0·150	0·145	0·141	0·141			
5·0		0·218	0·188	0·172	0·164	0·160	0·158			
5·5		0·250	0·215	0·194	0·184	0·178	0·176			
6·0		0·287	0·242	0·218	0·205	0·199	0·193			
6·5		0·330	0·279	0·246	0·228	0·219	0·213			
7·0		0·380	0·320	0·274	0·251	0·240	0·232			
7·5			0·368	0·305	0·276	0·263	0·252			
8·0			0·412	0·340	0·303	0·286	0·272			
8·5			0·474	0·377	0·332	0·310	0·294			
9·0			0·540	0·422	0·358	0·334	0·316			
9·5			0·620	0·471	0·398	0·359	0·338			
10·0			0·712	0·527	0·437	0·387	0·363			
10·5				0·591	0·481	0·416	0·388			
11·0				0·662	0·526	0·444	0·413			
11·5				0·753	0·582	0·472	0·441			
12·0				0·874	0·644	0·519	0·468			
12·5					0·715	0·549	0·500			
13·0					0·794	0·588	0·536			
13·5					0·892	0·636	0·574			
14·0					1·009	0·690	0·614			
14·5					1·180	0·758	0·658			
15·0					1·476	0·837	0·707			
15·5						0·935	0·761			
16·0						1·042	0·815			
16·5						1·198	0·884			
17·0						1·395	0·962			
17·5						1·678	1·050			
18·0						2·108	1·165	0·885		
18·5							1·286	0·952		
19·0							1·444	1·028		
19·5							1·658	1·107		
20·0							1·930	1·208		
20·5							2·349	1·316		
21·0								1·449		
21·5								1·607		
22·0								1·802		
22·5								2·050		
23·0								2·373		
23·5								2·808	1·646	
24·0									1·818	
24·5									1·994	
25·0									2·203	
25·5									2·463	
26·0									2·817	
26·5									3·246	
27·0										1·998
27·5										2·225
28·0										2·534
28·5										2·861
29·0										3·254
29·5										3·791

Table 6 General tree assortment table for conifers
Volume (OB) to 24 cm top diameter under bark

Length in metres	Height class in metres (height to 7 cm in brackets)								
	12 (9·5)	15 (12)	18 (15)	21 (18)	24 (21)	27 (24)	30 (27)	33 (30)	36 (33)
2·0	0·111	0·109	0·109	0·105	0·101	0·101	0·101		
2·5	0·150	0·144	0·138	0·133	0·132	0·130	0·128		
3·0	0·191	0·177	0·169	0·164	0·159	0·157	0·155		
3·5	0·234	0·212	0·203	0·196	0·191	0·190	0·182		
4·0		0·250	0·238	0·227	0·221	0·221	0·210		
4·5		0·291	0·275	0·261	0·254	0·254	0·239		
5·0		0·330	0·317	0·299	0·289	0·287	0·269		
5·5		0·392	0·358	0·336	0·325	0·320	0·299		
6·0		0·447	0·407	0·385	0·359	0·354	0·331		
6·5		0·513	0·453	0·422	0·400	0·389	0·364		
7·0		0·581	0·508	0·471	0·440	0·428	0·398		
7·5		0·661	0·566	0·521	0·485	0·465	0·433		
8·0		0·743	0·627	0·577	0·532	0·510	0·472		
8·5			0·692	0·637	0·581	0·558	0·513		
9·0			0·774	0·697	0·628	0·599	0·546		
9·5			0·852	0·769	0·685	0·645	0·594		
10·0				0·845	0·745	0·693	0·638		
10·5				0·937	0·808	0·748	0·689		
11·0				1·034	0·874	0·809	0·737		
11·5				1·148	0·948	0·868	0·785		
12·0				1·276	1·033	0·923	0·842		
12·5				1·478	1·128	0·993	0·901		
13·0					1·229	1·065	0·962		
13·5					1·359	1·141	1·019		
14·0					1·497	1·219	1·084		
14·5					1·679	1·340	1·144		
15·0					1·866	1·427	1·221		
15·5					2·087	1·552	1·302	1·133	
16·0						1·692	1·386	1·208	
16·5						1·822	1·472	1·285	
17·0						1·979	1·571	1·358	1·234
17·5						2·210	1·703	1·433	1·286
18·0						2·459	1·822	1·539	1·359
18·5						2·788	1·946	1·641	1·432
19·0							2·109	1·735	1·509
19·5							2·317	1·845	1·587
20·0							2·514	1·968	1·680
20·5							2·813	2·109	1·763
21·0							3·150	2·258	1·863
21·5								2·412	1·987
22·0								2·600	2·116
22·5								2·813	2·239
23·0								3·022	2·355
23·5								3·318	2·500
24·0								3·600	2·678
24·5									2·852
25·0									3·018
25·5									3·254
26·0									3·484
26·5									3·813

Hand Peeling Pines and Douglas Fir
East England

1 Conditions
The times apply under the following conditions:

a. The peeling of random length poles, sawlogs and 2 metre butt lengths.

b. Fresh felled pieces peeled by the feller within seven days of felling.

c. Pieces that are reasonably straight, free from large knots or swellings.

d. The different conditions in summer of 'sap up' and in winter of 'sap down'.

e. Scots and Corsican pine and Douglas fir.

2 Job specification
The times are for the following work:

a. Removing the outer bark from the piece leaving a clean smooth surface.

b. Where applicable, stacking pieces neatly at rack side.

3 Tools and equipment (including safety equipment)
a. Sandvik peeler with spare blades.

b. Sharpening stone.

c. Peeling cuddy (horse).

d. Protective clothing viz:
Industrial gloves.
Boots with safety toecap.

4 Allowances

Included in the times

a. For contingencies and work other than that performed on individual pieces, eg sharpening peeler, walking to new site, etc $7\frac{1}{2}\%$ of the time actually spent on peeling.

b. For personal needs and rest $17\frac{1}{2}$–25% of the total working time according to size of butt, pole or log.

5 Method of using the Output Guide
Paragraph 6 gives the time per pole or piece and the time per cubic metre for sawlogs. Entry is made to the table according to species and for poles and pieces the sap condition.

6 Times for hand peeling

| Species | Sap condition | Standard minutes per piece or pole | | Standard minutes per m^3 |
		Pole lengths 4·5–9·2 m	2 m butt length	Sawlogs 3, 6 and 9 m lengths with 14 cm minimum top diameter
Scots pine and	Up	7·01	3·49	
Douglas fir	Down	7·78	3·88	75·44
Corsican pine	Up	6·66	3·32	
	Down	7·40	3·69	

7 Modifications and variations to the times
N.B. These modifications and variations are only to be applied when the prevailing conditions or job specification differ from those listed in paragraphs 1 and 2.

a. *Butt peeling of poles for protection*
Where butt peeling of poles for protection is carried out, allow one-third of the normal standard minutes per pole.

Cundey Tractor-Mounted Splitting Saw

1 Conditions

The Output Guide applies to the splitting of pitprops under the following conditions:

a. Working on the forest road (see para 7a for depot working).

b. Props stacked by sizes in windrows, one side of the road and at right angles to the road. (Larger props—2·7 m in length—may have to be stacked parallel to the road with all small ends facing in the same direction.)

c. Props reasonably straight and free from large knots or swellings.

d. Prop stacks to be not higher than 1·5 m.

e. Props exceeding 130 mm top diameter to be split on a separate run to those under 130 mm top diameter as the infeed must be regulated to run at a slower speed and the rollers on the splitter require adjustment for the larger sizes.

f. Splitter equipped with 762 mm inserted tooth saw blade.

g. Splitting spruce and larch.

h. Two-man working for all sizes and exceptionally three-man working for props 2·70 m long.

2 Job specification

The Output Guide applies to the following work:

a. Splitting pitprops of the sizes shown in paragraph 6.

b. Props to be fed through splitter small end first.

c. The split props to be stacked neatly by size classes in windrows at roadside.

d. The number of splits by size classes to be counted and recorded.

e. Sawdust to be removed from the road and kept clear of drains.

f. Size of splits to be marked on stacks.

3 Tools and equipment (including safety equipment)

   a. Cundey splitting saw mounted on the three point linkage of a Ford 4000 or similar tractor.

   b. Tractor fuel, oil etc and all necessary tools and spares.

   c. Felt markers or crayons, notebook and pencils.

   d. Tarpaulin sheet (approximately $2\,m^2$ with a rope handle on one side) and shovel for removing sawdust.

   e. Wooden hammer and wedge.

   f. Protective clothing viz:
Safety helmet BS 2826, 5240.
Eye protection.
Ear defenders.
Industrial gloves.
Apron.
Boots with safety toecap.

4 Allowances

The following allowances are included in the Output Guide:

   a. For contingencies and work other than that performed on individual props and moving between stacks, eg refuelling, daily maintenance of tractor and splitter, removing sawdust from road, counting and recording number of pieces, marking stacks etc 24% of the time spent splitting and moving between stacks.

   b. For personal needs and rest 21%–46% of the total working time.

5 Method of using the Output Guide

   a. Ascertain the size of prop and the average number of props per stack. (Also number of men in team for props with lengths of $2\cdot7\,m$).

   b. Select the appropriate table from paragraph 6 and read off either output or standard minutes.

N.B. This information is given for either a two or three-man working team. Where the output per man is required the figures shown in paragraph 6 must be divided by the number of men in the team (ie two or three).

Where the machine is required the time per prop must be divided by the number of men in the team (ie two or three).

Example 1
Prop size 150 mm × 2·70 m
75 props per stack
Props carried 2·5 m
Two-man team
Time per prop = 2·24 standard minutes (from 6A)

Machine time $= \dfrac{2·24}{2} = 1·12$ standard minutes

Output per hour = 54 props (from 6B)

Output per man per hour $= \dfrac{54}{2} = 27$ props.

Example 2
Prop size 150 mm × 2·70 m
75 props per stack
Props carried 2·5 m
Three-man team
Time per prop = 1·71 standard minutes (from 6C)

Machine time $= \dfrac{1·71}{3} = 0·57$ standard minutes

Output per hour = 105 props (from 6D)

Output per man per hour $= \dfrac{105}{3} = 35$ props.

6A   Standard minutes per prop split (two men)
see page 302 overleaf

6B   Output in props split per hour by a two-man team
see page 303 overleaf

6C   Standard minutes per prop split (three men)

Number of props per stack	Prop size			
	150 mm	*160 mm*	*170 mm*	*180 mm*
	2·70 m	*2·70 m*	*2·70 m*	*2·70 m*
100	1·70	1·94	2·18	2·42
75	1·71	1·95	2·19	2·43
50	1·74	1·98	2·22	2·46
40	1·77	2·01	2·25	2·49
30	1·80	2.04	2·28	2·52
25	1·83	2·07	2·31	2·55
20	1·88	2·12	2·36	2·60
15	1·96	2·20	2·44	2·68
10	2·11	2·35	2·59	2·83

N.B. These times are for a three-man team, see paragraph 5.

6A Standard minutes per prop split (two men)

Number of props per stack	Prop size										
	100 mm	110 mm	110 mm	120 mm	150 mm	160 mm	160 mm	150 mm	160 mm	170 mm	180 mm
	1·05 m	1·05 m	1·35 m	1·35 m	1·80 m	1·80 m	1·95 m	2·70 m	2·70 m	2·70 m	2·70 m
150	0·31	0·35	0·43	0·47	—	—	—	—	—	—	—
100	0·32	0·36	0·44	0·48	0·95	0·99	1·15	2·23	2·57	2·87	3·19
75	0·33	0·37	0·45	0·49	0·96	1·00	1·16	2·24	2·58	2·88	3·20
50	0·35	0·39	0·47	0·51	0·98	1·02	1·18	2·26	2·60	2·90	3·22
40	0·37	0·41	0·48	0·52	1·00	1·04	1·20	2·28	2·62	2·92	3·24
30	0·39	0·43	0·51	0·55	1·02	1·06	1·22	2·30	2·64	2·94	3·26
25	0·41	0·45	0·53	0·57	1·04	1·08	1·24	2·32	2·66	2·96	3·28
20	0·44	0·48	0·56	0·60	1·07	1·11	1·27	2·35	2·69	2·99	3·31
15	0·49	0·53	0·61	0·65	1·13	1·17	1·33	2·40	2·74	3·04	3·36
10	0·60	0·64	0·71	0·75	1·23	1·27	1·43	2·51	2·85	3·15	3·47

N.B. These times are for a two-man team, see paragraph 5.

6B Output in props split per hour by a two-man team

Number of props per stack	Prop size										
	100 mm	110 mm	110 mm	120 mm	150 mm	160 mm	160 mm	150 mm	160 mm	170 mm	180 mm
	1·05 m	1·05 m	1·35 m	1·35 m	1·80 m	1·80 m	1·95 m	2·70 m	2·70 m	2·70 m	2·70 m
150	387	343	279	255	—	—	—	—	—	—	—
100	375	333	273	250	126	121	104	54	47	42	38
75	364	324	267	245	125	120	103	54	47	42	37
50	343	308	255	235	122	118	102	53	46	41	37
40	324	293	250	231	120	115	100	53	46	41	37
30	308	279	235	218	118	113	98	52	45	41	37
25	293	267	226	211	115	111	97	52	45	40	37
20	273	250	214	200	112	108	94	51	45	40	37
15	245	226	197	185	106	103	90	50	44	39	36
10	200	188	169	160	98	94	84	48	42	38	35

N.B. These outputs are for a two-man team, see paragraph 5.

6D   Output in props split per hour by a three-man team

Number of props per stack	Prop size			
	150 mm	160 mm	170 mm	180 mm
	2·70 m	2·70 m	2·70 m	2·70 m
100	106	93	83	74
75	105	92	82	74
50	103	91	81	73
40	102	90	80	72
30	100	88	79	71
25	98	87	78	70
20	96	85	76	69
15	92	82	74	67
10	85	77	69	64

N.B. These outputs are for a three-man team, see paragraph 5.

7   Modifications and variations to the Output Guide

N.B. These modifications and variations are only to be applied when the prevailing conditions or job specification differ from those listed in paragraphs 1 and 2.

a. *Depot working*

Where splitting takes place in a depot with a large concentration of material present the following times and outputs will apply.

	Prop size						
	100 mm	110 mm	110 mm	120 mm	150 mm	160 mm	160 mm
	1·05 m	1·05 m	1·35 m	1·35 m	1·80 m	1·80 m	1·95 m
SMs per prop	0·29	0·33	0·41	0·45	0·92	0·96	1·12
Output in props per hour	414	364	293	267	130	125	107

b. *Extra carrying of props*

Where props of 2·70 m long are carried more than 4 metres from the machine the expected output should be reduced as follows:

Paragraph 6B—Two-man team

        Props carried 4–5·99 metres—reduce output by 5 props per hour

        Props carried 6–9 metres—reduce output by 10 props per hour

Paragraph 6D—Three-man team

>Props carried 4–5·99 metres—reduce output by 10 props per hour
>
>Props carried 6–9 metres—reduce output by 20 props per hour

Paragraphs 6A and 6C should be modified using the following formula:

$$\text{Time per prop} = \frac{60}{\text{Output in props per hour}} \times \text{number of men in team.}$$

Example

Prop size 150 mm × 2·70 m

75 props per stack

Props carried 5 metres

Two-man team

Output per hour = 54 props (from paragraph 6B) minus 5 props (from paragraph 7b)

$\qquad\qquad\quad$ = 49 props

$\text{Time per prop} \quad = \dfrac{60}{49} \times 2 = 2\cdot45$ standard minutes.

Standard Times for Extraction by Massey Ferguson MF135 Fitted with Hydratongs

The tractor may be extracting produce from a first row thinning, from a selective thinning or from a clear felling and tables are given for each type of job.

A. Extraction from a first row thinning

1 Conditions

The standard times apply to:

a. The extraction of whole trees *or* converted produce from stump to ride. With whole trees no conversion has taken place, the stem being run out to about 3 cm top. Converted produce is in 2 m lengths with top diameters in two classes and tops/small poles. See Appendix for product descriptions.

b. Whole trees should lay with their butts facing the direction of extraction and centrally between the two adjoining rows. Converted produce should be hand piled at the side and parallel to the line, tops/small poles with butts in direction of extraction. Converted produce should not be set back in bays in the standing rows.

c. Rough tops and large branches have been pushed into intermediate rows by the feller.

d. Normal floor conditions, ie, flat or gently sloping ground up to 5° (9%), the planting furrows, new stumps and small lop and top being the only obstructions.

2 Job specification

The standard times are for the following work:

a. All produce to be extracted and stacked as and where directed. Tops/small poles and 2 metre lengths stacked to about 1 metre high. Whole trees to be left for conversion which may be in wide spread or compactly placed tushes. The heaps should be free of large masses of brash although small branches will inevitably be present.

b. Whole trees will be laying centrally in the line with the butt ends pointing in the direction of extraction. Tractor reverses into the line and by shunting some trees back and some forward makes small heaps—say three heaps of four. Then picking up the heap furthest in, drives forward to second, picks second and drives to first, picks first and drives out to dropping place. It will sometimes be possible to reverse over enough trees which can be picked up to make a full tush in one run forward. Two metre lengths will be hand piled at the side of, and parallel to, the line. The tractor is reversed into the line by-passing enough heaps to make up a full tush, then driven out of the line picking up heaps to make a full tush. Tops/smalls are treated in a similar way.

c. See also FC Leaflet No 55.

3 Tools and equipment (including safety equipment)

a. Massey Ferguson 135 fully guarded, fitted with safety cab and equipped with hydratongs.

b. Protective clothing viz:
Safety helmet BS 2826 or 5240.
Boots with safety toecap.
Ear defenders.

c. Tools to carry out daily maintenance and small repairs such as changing hydraulic lines and tightening nuts and bolts.

d. Tally counters for checking produce extracted.

e. For extracting 2 metre lengths a detachable cradle is fitted to the hydratong jib.

4 Allowances

Included in the standard times

a. For contingencies and work other than extracting trees and 2 metre lengths such as refuelling, greasing the grapple and linkage, checking oil and water, removing sticks from the tractor, removing brash from heaps of produce, getting instructions, making small adjustments to the tractor and grapple of up to 15 min, an allowance of 10% is made.

This does *not* include travelling to and from garage to work site or repairs/adjustments of over 15 minutes.

b. For personal needs and rest an allowance of 20% is made.

To be included in the price per standard minute

c. The appropriate incentive allowance.

5 Method of selecting the standard time

a. Table 6(i) gives terminal time per piece.

Table 6(ii) gives travelling time per piece in the line. To use this it will be necessary to calculate the average length of line to be travelled (usually $\frac{1}{2}$ or $\frac{1}{4}$ of the length of the compartment) and apply this distance to the table which gives the time for the return journey.

Table 6(iii) gives travelling time per piece on the rack and ride. To use this it will be necessary to calculate the average distance to be run and apply this distance to the table which gives times for the return journey.

Table 6(iv) gives travelling time to/from work site.

b. To arrive at a time for the job add together the appropriate times from the four tables.

c. In the context of these tables, 'terminal' refers to the making up of the tushes plus any shunting up and down the line involved plus dropping and stacking the tushes on the ride.

d. The times are calculated on the basis of the average number of pieces to a grapple and with the average distribution of produce as shown below.

Product	Pieces to a grapple	Heaps to a grapple	Distances to make a grapple full
Whole tree	9·09	Scattered	15·3 metres
2m, over 14cms TD	4·93	2·2	18·4 metres
2m, under 14cms TD	10·38	1·3	4·7 metres
Tops/smalls	20·50	2·5	45·2 metres

6 Tables of standard times for extraction from row thinning

Table (i) Terminal time per piece

Product	SMs per piece
Whole tree	0·45
2m length TD over 14cm	0·27
2m length TD under 14cm	0·10
Tops/smalls	0·10

Table (ii) Travelling time per piece in the row

Distance	Whole tree	2m TD over 14 cm	2m TD under 14 cm	Tops/smalls
20 m	0·08	0·16	0·09	0·03
30 m	0·12	0·21	0·12	0·04
40 m	0·15	0·27	0·15	0·05
50 m	0·19	0·32	0·18	0·06
60 m	0·22	0·37	0·21	0·07

Table (iii) Travelling time per piece on rack and/or ride

Distance	Whole tree	2m TD over 14 cm	2m TD under 14 cm	Tops/smalls
10 m	0·06	0·09	0·04	0·03
20 m	0·08	0·11	0·06	0·03
30 m	0·09	0·14	0·07	0·04
40 m	0·11	0·16	0·08	0·05
50 m	0·12	0·18	0·09	0·06
60 m	0·14	0·21	0·10	0·06
70 m	0·16	0·23	0·11	0·07
80 m	0·17	0·26	0·13	0·08
90 m	0·19	0·28	0·14	0·08
100 m	0·20	0·30	0·15	0·09
110 m	0·22	0·33	0·16	0·10
120 m	0·24	0·35	0·17	0·11
130 m	0·25	0·37	0·19	0·11
140 m	0·27	0·40	0·20	0·12
150 m	0·28	0·42	0·21	0·13
160 m	0·30	0·45	0·22	0·13
170 m	0·32	0·47	0·23	0·14
180 m	0·33	0·49	0·24	0·15
190 m	0·35	0·52	0·26	0·16
200 m	0·36	0·54	0·27	0·16

Table (iv) Travelling to/from work site

For travelling to and from work site on public and forest road add 7·2 SMs per mile.

B. Extraction from a selective thinning

1 Conditions
The standard times apply to:
a. The extraction of 'poles', 'multiples' and 'logs' from stump to ride (see Appendix for description of these products).

b. The majority of products laying with their butts pointing in the direction of extraction.

c. 100% brashing.

d. Normal floor conditions, ie, flat or gently sloping ground, a fair amount of tops, brash and vegetation, stumps left from the thinnings, no deep drains but shallow furrows from planting still evident.

e. 650 to 2,150 stems per hectare still standing.

f. Racks about 40 metres apart.

2 Job specification

The standard times are for the following work:

a. All products to be extracted, sorted and stacked as and where directed for conversion or loading. Poles to be left for conversion in tushes which may be compact or spread. Multiples and logs to be stacked two tushes high to about 1 m. Poles may be sorted into straight and crooked, multiples into graded lengths and logs into size categories.

b. The job divides into two stages: (i) the stump/rack stage and (ii) the rack/ride stage. Stage one is best accomplished by reversing into the wood and shunting products into small sorted heaps then driving out to the rack accumulating as many pieces of one product as possible en route. Where any one product is scarce it should be brought out to the rack in mixture and every effort should be made to ensure that as many pieces as possible are carried on any journey. Stage two is the picking up of the sorted heaps to make a grapple full or tush and pulling the tush down rack and ride to stack it where and as directed. Stacking may be accomplished with hydratongs by turning the tractor at right angles to the tush while still gripping it and then reversing on to the heap of produce, releasing the butt end of the tush and then reversing the tip end on to the heap in a similar way. To avoid damage to the grapple the driver should ensure that the tush is released before driving away.

3 Tools and equipment

As in row thinning (Part A) but no 2 metre cradle necessary.

4 Allowances

As in row thinning (Part A).

5 Method of selecting the standard time

The times are divided into four tables.

a. Table 6(i) gives time per piece for stage one. This is the time needed to get produce from stump to rack (including the travelling on the rack and in the wood and the necessary shunting) and leaving it in sorted heaps.

Table 6(ii) gives terminal time per tush for stage two. This is the time needed to make up a full tush of sorted produce on the rack and drop it and stack it on the ride. To use this table it will be necessary to know the average number of pieces that make up a tush in order to arrive at a time per piece. The number of pieces given are the average encountered in studies.

Table 6(iii) gives travelling time per tush for rack and ride. To use this table it is necessary to know the average number of pieces that make up a tush in order to arrive at a time per piece. It is also necessary to calculate the average distance to be travelled on rack and ride. (Usually $\frac{1}{2}$ or $\frac{1}{4}$ of the length of the compartment for rack travel and dependent on directions for ride travel.)

Table 6(iv) gives travelling time to/from work site.

b. To arrive at a time for the whole job add together the times derived from the four tables. If hydratongs are performing only stage two, add together the times derived from tables 6(ii), 6(iii) and 6(iv).

6 Standard times for extracting produce from a thinning

Table (i) Times for stage one

Product	SMs per piece
Poles	0·96
Multiples	1·14
Logs	1·19

Table (ii) Terminal times for stage two

Product	SMs per tush	Pieces per tush
Poles	0·66	8·0
Multiples	1·15	5·1
Logs	1·35	2·8

Table (iii) Travelling times on rack and ride

Distance	SMs per tush	Distance	SMs per tush
10 m	0·56	160 m	2·77
20 m	0·71	170 m	2·92
30 m	0·85	180 m	3·06
40 m	1·00	190 m	3·21
50 m	1·15	200 m	3·36
60 m	1·30	220 m	3·65
70 m	1·44	240 m	3·95
80 m	1·59	260 m	4·24
90 m	1·74	280 m	4·54
100 m	1·89	300 m	4·83
110 m	2·03	320 m	5·13
120 m	2·18	340 m	5·42
130 m	2·33	360 m	5·72
140 m	2·48	380 m	6·01
150 m	2·62	400 m	6·30

Table (iv) Travelling to/from work site

For travelling to and from work site on public and forest road add 7·2 SMs per mile.

C. Extraction from a clearfelling

1 Conditions

The standard times apply to:

a. The extraction of 'poles', 'multiples' and 'logs' from stump to ride. The felling in question is that where a feller has two blocks in one of which he is felling while extraction takes place from the other. See Appendix for descriptions of the products.

b. Felling has been done at an angle to the face of the block so that the majority of the produce lays with the butt ends inclined at an angle towards the extraction rack.

c. Normal floor conditions, ie, flat or gently sloping ground, a fair amount of lop and top and vegetation, stumps left from previous thinnings and the felling but few coppice stumps, no deep drains.

d. Racks 35–40 metres apart.

e. Logs are graded at some point before stacking takes place. The times do not allow for pulling the rideside heaps of logs to pieces in order to sort and re-stack them.

2 Job specification

a. All produce to be extracted, sorted and stacked as and where directed. Poles are usually left for conversion in compact or spread tushes, while multiples and logs are stacked two tushes high (about 1 m) to await collection. Sorting of the poles may be into straight and crooked, multiples into graded lengths and logs into volume categories with additional sorting for straight or crooked.

b. The job divides into two stages, stump to rack and rack to ride. The first stage must be accomplished before the feller returns from his other block. The recommended method is as in thinnings. Logs are often measured and marked while they lay at the end of stage one and they may be graded as something other than the tractor driver has anticipated. Stage two therefore includes some sortings of logs, making small heaps into full tushes, picking up, pulling out and stacking produce as and where directed.

3 Tools and equipment
As for row thinning (Part A) but no 2 metre cradle necessary.

4 Allowances
As in row thinning (Part A).

5 Method of selecting the standard times
a. The times are divided into four tables.

Table 6(i) gives the time per piece for stage one. This is the time needed to get the product from stump to rack including movement on the rack and in the wood, the necessary shunting and dropping into sorted heaps.

Table 6(ii) gives the terminal time per tush for stage two. This is the time needed to make up a full tush of sorted produce on the rack and drop it and stack it on the ride. It is given in two parts, one where stage one has been performed by the hydratongs and the other where stage one has been performed by Thetford tongs. To arrive at a time per piece it will be necessary to know the average number of pieces that make up a tush. The numbers of pieces given are those encountered in studies.

Table 6(iii) gives travelling time per tush in rack and ride. It will again be necessary to know the average number of pieces to a tush in order to calculate a time per piece. It will also be necessary to calculate the average distance to be travelled on rack and ride. (Rack distance is usually $\frac{1}{2}$ or $\frac{1}{4}$ the length of the compartment. Ride distance will depend on the directions given the tractor driver.)

Table 6(iv) gives the travelling time to/from work site. To arrive at a time for the whole job add together the times derived from the four tables.

b. If the hydratongs are performing just stage two then only tables 6(ii), 6(iii) and 6(iv) are used.

6 Standard times for extracting produce from a clearfelling

Table (i) Times for stage one

Product	SMs per piece
Poles	0·91
Multiples	1·09
Logs	0·86

Table (ii) Terminal times for stage two

Product	Time per tush if stage 1 by hydratong	Time per tush if stage 1 by Thetford tong	Pieces per tush
Poles	0·67	2·22	7·3
Multiples	1·53	2·72	5·2
Logs	1·42	2·39	2·4

0·87 SMs of the multiples and logs time is for stacking.
0·16 SMs of the logs time is for sorting after grading.

Table (iii) Travelling time on rack and ride

Distance	SMs per tush	Distance	SMs per tush
10 m	0·56	160	2·77
20 m	0·71	170	2·92
30 m	0·85	180	3·06
40 m	1·00	190	3·21
50 m	1·15	200	3·36
60 m	1·30	220	3·65
70 m	1·44	240	3·95
80 m	1·59	260	4·24
90 m	1·74	280	4·54
100 m	1·89	300	4·83
110 m	2·03	320	5·13
120 m	2·18	340	5·42
130 m	2·33	360	5·72
140 m	2·48	380	6·01
150 m	2·62	400	6·30

Table (iv) Travelling to/from work site

For travelling to and from work site on public and forest road add 7·2 SMs per mile.

Product measurements
Throughout the tables reference is made to poles, multiples and logs in thinnings and fellings and to whole trees, 2 metre lengths and tops/smalls in line thinnings. A description of these products and some measurements follow:

Pole
The upper part of the stem left when a log and/or multiple has been removed. The butt diameter is about 16 cm and the length about 6 m.

Multiple
Usually the middle part of the stem. A log has been cut from the bottom and a pole from the top. The length is a multiple of 2 metres usually 4 m or 6 m and the butt diameter about 22 cm.

Logs
Usually the bottom part of the tree but may be a second or third cut. The butt diameter is about 30 cm and the length from 3·3 to 6·6 m.

2 metre lengths
Trees in a line thinning converted into 2 metre lengths to a minimum top diameter of 7 cm. Butt diameters are 17 or 13 cm in the top diameter classes of over 14 and under 14 cm respectively.

Tops/smalls
Trees or parts of trees in a line thinning from which a 2 metre length can not be cut. The butt diameter is about 9 cm and the length about 5·3 m.

Whole trees
Trees in a line thinning where no conversion has taken place. Stem left whole to about 3 cm top diameter. The butt diameter is about 16 cm and the length about 8·7 m.

Pole Length Ground Skidding by County Super-Four and Igland Winch
Southern England

1 Conditions

The times apply to pole lengths extracted under the following conditions:

a. Thinning or clearfelling.

b. Poles felled directionally as instructed by the supervisor.

c. The average tree volume determined by an agreed method.

d. Drains of 1 m or more wide and 0·75 m or more deep to be filled in with brash, tops, etc so as to be readily traversable.

e. Rack espacement—40 m or less (thinning only), with adequate turning space provided in the rack every 50 m in length.

f. Ample room for stacking poles and manoeuvring at roadside/ conversion bay/lower landing.

g. Chokering and unchokering carried out by the tractor driver.

2 Job specification

The times apply to the following work:

a. Tractor manoeuvred to face extraction direction, in rack or wood, prior to b. below.

b. Winch rope hauled out and chokers distributed to butt or tip of each pole to be extracted.

c. Poles chokered and secured to winch rope working back to the tractor from the tail pole.

d. Repeat b. and c. above with the second winch rope, if necessary after further manoeuvring of the tractor.

e. Load winched to butt plate and then lifted as high as possible from the ground.

f. Skid load to stacking area.

g. Unchoker load, wind in winch ropes and secure chokers to the tractor.

h. Return to extract the next load.

i. Normal daily maintenance of the tractor, winch and chokers.

j. Stack timber at roadside as required.

3 Tools and equipment (including safety equipment)
County Super-four tractor (654 or 754) equipped with:

a. Log rolling blade.

b. Igland Spesial 4 tonne double drum winch and ropes plus sufficient (and spare) keyhole sliders and chain chokers or EIA hooks and polypropylene chokers.

c. The appropriate maintenance tools.

d. Protective clothing viz:
Safety helmet BS 2826 or 5240.
Ear defenders.
Gloves with wireproof palms.
Boots with safety toecap.

4 Allowances
Included in the times

a. For work other than that of actual extraction (terminal and movement) 20% of the time spent extracting.

b. For personal needs and rest 20% of the total working time.

5 Method of selecting the time
Entry to the tables in paragraph 6 is gained by using the average volume of the poles to be extracted, the average extraction distance, the difficulty of the terrain and whether skidding is from a thinning or clearfelling.

The terrain conditions should be assessed on site and the appropriate points allocated from the tables shown below. The total points awarded from these tables are then used to fix the terrain classification in table 5B.

A   Terrain score tables for wheeled tractor extraction
Points should be allocated from tables (i), (ii) or (iii) and (iv) and totalled for use in table 5B.

Table (i)   Slope

Slope in percentage	Slope in degrees	Points for uphill extraction	Points for downhill extraction
0	0	3	3
2	1	4	2
3	2	6	1
5	3	8	0
7	4	10	0
9	5	12	1
10	6	14	2
12	7	16	4
14	8	18	6
16	9	21	9
18	10	24	12
19	11	27	15
21	12	30	18
23	13	Not considered	21
25	14	suitable for this	24
27	15	type of extraction	27
29	16		30

Table (ii)   Soil condition
Where felling litter is present in such quantities that it makes an assessment of the soil type difficult use the table shown below. In all other cases use table (iii). Soil moisture refers to conditions at the time of the operation.

Soil moisture	Points
Dry	0–5
Moist	6–10
Wet	11–15

Table (iii)   Soil condition and type

Soil moisture	Soil type (15–30 cms below surface)				
	Gravel	Loam	Clay	Peat	Sand
Dry	0	0	0	0	7
Moist	4	8	12	18	10
Wet	4	15	21	31	19

Table (iv)   Impediments
Ditches, stumps, rocks, etc.

Impediments	Points
Few	0–5
Moderate	6–10
Severe	11–15
Very severe	16–20

B   *Terrain classification for County Super-four equipped with double drum winch and log rolling blade*

Total points awarded in paragraph 5A	Terrain classification
0–10	Easy—turn to paragraph 6A
11–20	Moderate—turn to paragraph 6B
21–30	Difficult—turn to paragraph 6C
31 +	Very difficult—not considered suitable for the County Super-four

Example

The site:
Slope 5° (9%)
Uphill extraction
Average extraction distance 300 metres
Little felling litter over moist clay
Few impediments

The crop:
P. 48, Douglas fir, average volume of pole removed $0.20\,\text{m}^3$, thinning
From table 5A (i)   12 points
From table 5A (iii) 12 points
From table 5A (iv)   3 points
Total = 27 points
Terrain classification from table 5B—difficult
Time per pole from paragraph 6C(i)—6.5 minutes
Output in poles per hour from paragraph 6C(i)—9

6 Time for extraction in minutes per pole

A *Easy terrain (total score 0–10)*

(i) *Thinning*

Average pole volume in m³	Average extraction distance in metres					Average number of poles per load
	100	200	300	400	500	
0·05	1·9	2·0	2·1	2·2	2·3	20
0·10	2·8	3·1	3·3	3·5	3·6	15
0·15	3·6	3·9	4·2	4·4	4·6	13
0·20	4·0	4·4	4·7	5·1	5·3	12
0·25	4·3	4·7	5·1	5·5	5·9	10
0·30	4·6	5·1	5·6	6·0	6·4	8
0·40	5·3	5·8	6·4	7·0	7·6	6
0·50	5·9	6·5	7·2	7·9	8·7	5
0·60	6·5	7·2	7·9	8·6	9·6	4
0·70	7·3	8·0	8·9	9·8	10·8	3–4

(ii) *Clearfelling*

Average pole volume in m³	Average extraction distance in metres					Average number of poles per load
	100	200	300	400	500	
0·20	2·7	3·1	3·5	3·8	4·1	12
0·30	3·6	4·1	4·6	5·1	5·6	8
0·40	4·5	5·1	5·8	6·5	7·2	6
0·50	5·3	6·1	6·9	7·9	8·6	5
0·60	5·7	6·6	7·7	8·8	9·8	4
0·70	5·9	6·8	8·1	9·5	10·7	3–4
0·80	6·0	7·2	8·5	9·9	11·5	3
0·90	6·1	7·4	8·8	10·1	11·7	3
1·00	6·2	7·6	9·1	10·5	12·1	2–3
1·50	7·5	8·6	10·0	12·0	15·0	2
2·00	10·0	10·9	12·0	15·0	20·0	1–2

6 Output in number of poles extracted per hour

A *Easy terrain (total score 0–10)*

(i) *Thinning*

Average pole volume in m³	Average extraction distance in metres					Average number of poles per load
	100	*200*	*300*	*400*	*500*	
0·05	32	31	30	29	28	20
0·10	21	19	18	17	16	15
0·15	17	16	15	14	13	13
0·20	15	14	13	12	11	12
0·25	14	13	12	11	10	10
0·30	13	12	11	10	9	8
0·40	11	10	9	$8\frac{1}{2}$	8	6
0·50	10	9	8	$7\frac{1}{2}$	7	5
0·60	9	8	$7\frac{1}{2}$	7	6	4
0·70	$8\frac{1}{2}$	$7\frac{1}{2}$	7	6	$5\frac{1}{2}$	3–4

(ii) *Clearfelling*

Average pole volume in m³	Average extraction distance in metres					Average number of poles per load
	100	*200*	*300*	*400*	*500*	
0·20	23	20	18	16	15	12
0·30	17	15	13	$11\frac{1}{2}$	10	8
0·40	14	12	10	9	8	6
0·50	12	10	9	8	7	5
0·60	11	$9\frac{1}{2}$	8	7	6	4
0·70	$10\frac{1}{2}$	9	$7\frac{1}{2}$	$6\frac{1}{2}$	$5\frac{1}{2}$	3–4
0·80	$10\frac{1}{2}$	9	$7\frac{1}{2}$	6	5	3
0·90	10	9	7	6	5	3
1·00	10	9	7	6	5	2–3
1·50	8	7	6	5	4	2
2·00	6	$5\frac{1}{2}$	5	4	3	1–2

6 Time for extraction in minutes per pole

B *Moderate terrain (total score 11–20)*

(i) *Thinning*

Average pole volume in m³	Average extraction distance in metres					Average number of poles per load
	100	200	300	400	500	
0·05	2·0	2·2	2·4	2·6	2·8	20
0·10	3·2	3·4	3·6	3·9	4·1	15
0·15	3·7	4·2	4·5	4·9	5·1	13
0·20	4·4	4·8	5·2	5·7	6·1	12
0·25	4·9	5·4	5·8	6·4	6·8	10
0·30	5·3	5·8	6·4	7·0	7·7	8
0·40	5·7	6·5	7·3	8·3	9·4	6
0·50	6·2	7·4	8·6	9·8	11·2	5
0·60	6·9	8·2	9·5	11·0	12·7	4
0·70	7·6	9·0	10·5	12·2	14·3	3–4

(ii) *Clearfelling*

Average pole volume in m³	Average extraction distance in metres					Average number of poles per load
	100	200	300	400	500	
0·20	2·9	3·3	3·7	4·1	4·5	12
0·30	3·8	4·4	5·0	5·6	6·2	8
0·40	4·8	5·5	6·3	7·2	8·0	6
0·50	5·6	6·5	7·5	8·6	9·7	5
0·60	5·9	7·1	8·4	9·6	10·9	4
0·70	6·3	7·7	9·2	10·8	12·3	3
0·80	6·6	8·3	10·0	11·9	13·6	3
0·90	6·7	8·6	10·2	12·3	14·4	2–3
1·00	7·0	8·8	10·7	13·0	15·2	2
1·50	8·6	10·0	12·0	14·0	16·2	1–2
2·00	12·0	15·0	17·1	20·0	24·0	1–2

6 Output in number of poles extracted per hour

B *Moderate Terrain (total score 11–20)*

(i) *Thinning*

Average pole volume in m³	Average extraction distance in metres					Average number of poles per load
	100	*200*	*300*	*400*	*500*	
0·05	30	27	25	23	22	20
0·10	19	17	16	$15\frac{1}{2}$	15	15
0·15	16	14	13	$12\frac{1}{2}$	12	13
0·20	14	12	11	$10\frac{1}{2}$	10	12
0·25	12	11	10	$9\frac{1}{2}$	9	10
0·30	11	10	9	$8\frac{1}{2}$	8	8
0·40	10	9	8	7	$6\frac{1}{2}$	6
0·50	9	8	7	6	$5\frac{1}{2}$	5
0·60	$8\frac{1}{2}$	$7\frac{1}{2}$	$6\frac{1}{2}$	$5\frac{1}{2}$	5	4
0·70	8	7	6	5	4	3–4

(ii) *Clearfelling*

Average pole volume in m³	Average extraction distance in metres					Average number of poles per load
	100	*200*	*300*	*400*	*500*	
0·20	22	19	17	15	13	12
0·30	15	$13\frac{1}{2}$	12	10	9	8
0·40	13	11	9	8	7	6
0·50	11	$9\frac{1}{2}$	8	7	6	5
0·60	$10\frac{1}{2}$	9	7	6	5	4
0·70	10	8	$6\frac{1}{2}$	$5\frac{1}{2}$	$4\frac{1}{2}$	3
0·80	$9\frac{1}{2}$	$7\frac{1}{2}$	6	5	4	3
0·90	9	$7\frac{1}{2}$	6	5	4	2–3
1·00	9	$7\frac{1}{2}$	6	5	4	2
1·50	7	6	5	4	$3\frac{1}{2}$	1–2
2·00	5	4	$3\frac{1}{2}$	3	$2\frac{1}{2}$	1

6 Time for extraction in minutes per pole

C *Difficult terrain (total score 21–30)*

(i) *Thinning*

Average pole volume in m³	Average extraction distance in metres					Average number of poles per load
	100	*200*	*300*	*400*	*500*	
0·05	2·2	2·6	3·2	3·8	4·7	20
0·10	3·4	3·9	4·4	5·0	5·6	15
0·15	4·2	4·8	5·5	6·2	7·1	13
0·20	4·8	5·6	6·5	7·5	8·6	10
0·25	5·4	6·4	7·5	8·7	9·9	8
0·30	6·0	7·2	8·6	9·9	11·1	7
0·40	6·9	8·3	10·0	11·9	14·0	5–6
0·50	7·5	9·5	11·7	14·1	16·8	4–5
0·60	8·3	10·4	13·0	15·8	19·2	3–4
0·70	9·0	11·4	14·2	17·4	21·2	3

(ii) *Clearfelling*

Average pole volume in m³	Average extraction distance in metres					Average number of poles per load
	100	*200*	*300*	*400*	*500*	
0·20	3·3	4·2	5·2	6·2	7·1	10
0·30	4·6	5·9	7·2	8·6	9·8	7
0·40	5·6	7·3	9·2	11·2	12·6	5–6
0·50	6·5	8·6	10·7	13·3	15·4	4–5
0·60	7·2	9·9	12·2	15·3	17·8	3–4
0·70	7·8	10·7	13·7	16·8	19·3	3
0·80	8·0	11·5	15·0	17·8	21·1	2–3
0·90	8·6	12·3	16·0	19·4	22·5	2
1·00	9·3	13·0	17·1	20·7	23·5	1–2
1·50	12·0	15·0	20·0	24·0	30·0	1–2
2·00	15·0	20·0	24·0	30·0	40·0	1

6 Output in number of poles extracted per hour

C *Difficult terrain (total score 21–30)*

(i) *Thinning*

Average pole volume in m³	Average extraction distance in metres					Average number of poles per load
	100	*200*	*300*	*400*	*500*	
0·05	28	23	19	16	13	20
0·10	17	15	13	12	10	15
0·15	14	12	11	10	8	13
0·20	12	11	9	8	7	10
0·25	11	9	8	7	6	8
0·30	10	8	7	6	5	7
0·40	9	7	6	5	4	5–6
0·50	8	$6\frac{1}{2}$	5	$4\frac{1}{2}$	$3\frac{1}{2}$	4–5
0·60	7	6	$4\frac{1}{2}$	4	$3\frac{1}{2}$	3–4
0·70	$6\frac{1}{2}$	$5\frac{1}{2}$	4	$3\frac{1}{2}$	3	3

(ii) *Clearfelling*

Average pole volume in m³	Average extraction distance in metres					Average number of poles per load
	100	*200*	*300*	*400*	*500*	
0·20	20	15	12	10	9	10
0·30	14	10	8	7	6	7
0·40	11	9	7	$5\frac{1}{2}$	$4\frac{1}{2}$	5–6
0·50	10	8	6	$4\frac{1}{2}$	4	4–5
0·60	9	$6\frac{1}{2}$	5	4	$3\frac{1}{2}$	3–4
0·70	8	6	$4\frac{1}{2}$	$3\frac{1}{2}$	3	3
0·80	$7\frac{1}{2}$	$5\frac{1}{2}$	4	3	$3\frac{1}{2}$	2–3
0·90	7	5	4	3	$2\frac{1}{2}$	2
1·00	7	5	$3\frac{1}{2}$	3	$2\frac{1}{2}$	1–2
1·50	5	4	3	$2\frac{1}{2}$	2	1–2
2·00	4	3	$2\frac{1}{2}$	2	$1\frac{1}{2}$	1

7 Modifications
None.

Extraction of Logs, Pulpwood and Poles by High Lead Double Drum Winch

1 Conditions

The times apply to extraction under the following conditions:

a. Logs and pulpwood cut to Scottish pulp specification (length 3·0 m with tolerance down to 2·5 m; diameter 5–40 cm over bark).

b. Logs and points.

c. Poles.

d. Pulp loads to average 0·36 cubic metres and be so prepared and positioned by the fellers that one end is raised on a bearer to facilitate chokering and access to the rack is unimpeded.

e. Both uphill and downhill extraction to the road.

f. A team of two men.

g. Rack espacement of 23 metres, rack width of 3 metres minimum, end at a single suitable spar tree with natural anchor, single natural head anchor for tractor.

h. Adequate stacking space at roadside without excessive foundation building required.

i. Tractor in line with the rack.

j. Slopes of up to 40° (84%) with a fair amount of rocks, brash and tops which must not exceed 2·5 metres in length.

2 Job specification

The times apply to the following work:

a. All produce to be extracted and stacked neatly at roadside so as not to obstruct the passage of vehicles.

b. Stacks to be made at right angles to the road with ends flush for grapple loading.

c. Produce types to be stacked separately.

3 Tools and equipment (including safety equipment)

    a. Isachsen Mark III double drum winch mounted on a Massey Ferguson 35 or 135 model tractor.

    or

    b. Igland Spesial double drum winch mounted on a Ford 3000 or Massey Ferguson 135 tractor.

    N.B. This guide also applies to Ford 3000 Igland Spesial Skyline outfits when used as a high lead even though the haul-back cable is 7 mm instead of 8 mm diameter. Both a. and b. above should be equipped to Forestry Commission Mechanical Plant Specification 70/29.

    Additional equipment:
    Two radio pack sets, or one pack set with a mobile and loudspeaker at the tractor end.
    Ground anchors.
    Power saw.

    c. Protective clothing viz:
    Safety helmet BS 2826 or 5240.
    Ear defenders.
    Gloves with wireproof palms.
    Boots with safety toecap.
    Chokerman need not wear ear defenders.
    In addition, outer clothing must be devoid of any loose flaps likely to get caught in the moving parts of the machinery or running cables.

4 Allowances

Included in the times

    a. For contingencies and work other than extraction, eg refuelling, daily maintenance of the tractor and winch, to and from work sites, moving tractor to let road traffic pass, counting and checking, improving stacking sites, removal of obstructions from the rack, 16% of the time spent on actual set up and take down and extraction.

    b. For personal needs and rest 27% of the total working time.

5 Method of using the Output Guide

    a. The Output Guide is composed of separate tables for set up and take down, and for extraction. The following facts are required to select the time for each part:

328

1 The average length of rack measured from roadside to the foot of the spar tree.
2 The total volume coming from the area.
3 The average load size.
4 Type of winch in use ie Isachsen or Igland.

b. *Set up and take down*
The time per cubic metre for set up and take down is dependent on rack espacement (or numbers of racks in the area), yield of the area, and rack length. The steps to determine the time per cubic metre are:
1 Divide the number of racks into the tariff volume (see FC Booklet No 36)—volume per rack.
2 From the average rack length and table 6 obtain the time per rack.
3 Divide the time per rack from 2 above by the volume per rack, giving time per cubic metre for set up and take down.

c. *Extraction*
The time per cubic metre for extraction can be read directly from table 6 when the average rack length and load size are known.

d. *Total time for the job*
The time for the job is the sum of b.3 and c. above, ie time per cubic metre for set up and take down plus extraction time per cubic metre.

e. *Notes:*
1 The average length of rack can be measured by pacing on the ground the length of a sample of racks (eg every third). On broken, hillocky ground rack length will be very variable and most likely every rack will have to be measured. The easiest time to note the rack length is at the time of marking.
2 If the volume per hectare or the rack length varies appreciably from one end of the working area to another, break the area down into uniform blocks if possible, with a separate volume or rack length for each block.
3 The average load size is best assessed when going over the area after felling is complete and measuring sample loads. This can be checked by the winch operator recording the total number of loads extracted and dividing this into the total volume for the working area.

N.B. *The times shown in paragraph 6 apply to a two-man team.* For use in forecasting machine requirements the times should be halved, eg time for two men to set up and take down equals 86 standard minutes, machine time equals 86 over 2 equals 43 standard minutes.

6 Time per cubic metre for extraction by Isachsen Mark III and Igland Spesial double drum winches

A. *Set up and take down—Isachsen and Igland*

Rack length in metres	Standard minutes per rack for set up and take down
40	68
50	71
60	75
70	79
80	82
90	86
100	90
110	94
120	97
130	101
140	105

N.B. *These times are for a two-man team, see paragraph 5.*

B. *Time for extraction by Isachsen Mark III on MF35 or MF135*

Rack length in m	Standard minutes per m³ for extraction							
	Average load size m³							
	0·50	0·45	0·40	0·35	0·30	0·25	0·20	0·15
40	12·20	13·56	15·25	17·43	20·33	24·40	30·50	40·67
50	12·80	14·22	16·00	18·29	21·33	25·60	32·00	42·67
60	13·40	14·89	16·75	19·14	22·33	26·80	33·50	44·67
70	14·00	15·56	17·50	20·00	23·33	28·00	35·00	46·67
80	14·60	16·22	18·25	20·86	24·33	29·20	36·50	48·67
90	15·20	16·89	19·00	21·71	25·33	30·40	38·00	50·67
100	15·80	17·56	19·75	22·57	26·33	31·60	39·50	52·67
110	16·40	18·22	20·50	23·43	27·33	32·80	41·00	54·67
120	17·00	18·89	21·25	24·29	28·33	34·00	42·50	56·67
130	17·60	19·56	22·00	25·14	29·33	35·20	44·00	58·67
140	18·20	20·22	22·75	26·00	30·30	36·40	45·50	60·67

N.B. *These times are for a two-man team, see paragraph 5.*

C. *Time for extraction by Igland Spesial on Ford 3000 or MF135 tractor*

Rack length in m	Standard minutes per m³ for extraction							
	Average load size m³							
	0·50	0·45	0·40	0·35	0·30	0·25	0·20	0·15
40	11·96	13·29	14·95	17·09	19·73	23·92	29·90	39·87
50	12·48	13·87	15·60	17·83	20·80	24·96	31·20	41·60
60	13·12	14·58	16·40	18·74	21·87	26·24	32·80	43·73
70	13·68	15·20	17·10	19·54	22·80	27·36	34·20	45·60
80	14·28	15·87	17·85	20·40	23·80	28·56	35·70	47·60
90	14·84	16·49	18·55	21·20	24·73	29·68	37·10	49·47
100	15·40	17·11	19·25	22·00	25·67	30·80	38·50	51·33
110	15·96	17·73	19·95	22·80	26·60	31·92	39·90	53·20
120	16·56	18·40	20·70	23·66	27·60	33·12	41·40	55·20
130	17·12	19·02	21·40	24·46	28·53	34·24	42·80	57·07
140	17·72	19·69	22·15	25·31	29·53	35·44	44·30	59·07

N.B. *These times are for a two-man team, see paragraph 5.*

D. *Output in m³ extracted per hour, Isachsen Mark III winch on MF35 or MF135 tractor (N.B. excludes set up and take down)*

Rack length in m	Average load size in m³							
	0·50	0·45	0·40	0·35	0·30	0·25	0·20	0·15
40	9·836	8·850	7·869	6·885	5·903	4·918	3·934	2·951
50	9·375	8·439	7·500	6·561	5·626	4·688	3·750	2·812
60	8·955	8·059	7·164	6·270	5·374	4·478	3·582	2·686
70	8·571	7·712	6·857	6·000	5·144	4·286	3·429	2·571
80	8·219	7·398	6·575	5·753	4·932	4·110	3·288	2·466
90	7·895	7·105	6·316	5·527	4·737	3·947	3·158	2·368
100	7·595	6·834	6·076	5·317	4·558	3·797	3·038	2·278
110	7·317	6·586	5·854	5·122	4·391	3·659	2·927	2·195
120	7·059	6·353	5·647	4·940	4·236	3·529	2·824	2·118
130	6·818	6·135	5·455	4·773	4·091	3·409	2·727	2·045
140	6·593	5·935	5·275	4·615	3·960	3·297	2·637	1·978

7 Modifications and variations to the times and outputs

N.B. These modifications and variations are only to be applied when the prevailing conditions or job specification differ from those listed in paragraphs 1 and 2.

a. *Downhill extraction*

When racks are over 120 metres long and the slope is steeper than:

15° (27%), extraction all downhill—*add* 5% to the set up and take down time.

20° (36%), extraction all downhill—*add* up to 10% to the set up and take down time.

b. *Difficult stacking space*

The winch operator normally moves pieces of pulpwood up to 4·5 metres when stacking. Where, due to site limitation or volume of produce, greater movement distances are involved add the following which should be limited to the racks concerned:

Carrying over 4·5 metres but less than 9 metres—*add* 5% to the extraction time or reduce the output by 5%.

Carrying over 9 metres but less than 13·5 metres—*add* $7\frac{1}{2}$% to the extraction time or reduce the output by $7\frac{1}{2}$%.

c. *Trewhella ground anchors*

Where such anchors are required—*add* 12·0 standard minutes to the set up and take down time for each plate used.

d. *Spar tree substitute*

Where an A frame is used as a spar tree substitute—*add* 60 standard minutes to the time for set up and take down.

e. *Long splicing of ropes*

The time taken to long splice ropes is not included in the Output Guide and must be paid for separately.

Volvo SM 462 Forwarder and
Volvo SM 868 Forwarder/Grapple Skidder
South-West England

1 Conditions

The Output Guide applies to extraction under the following conditions:

a. Extraction of produce cut to the following specifications:

Product	Length in metres	Diameter over bark in centimetres
SABM	2·5–3	8–30
EC chipper	1·85	15–25
Woodcraft/Coates	1·85	15–30
EC chipper	2·46	13–23
Woodcemair	2·0	13–28
Sawlogs	3·66–8·23	16–21
Stakes	1·37	5–7 (TD)
Stakes	1·67	7–10 (TD)

b. Extraction of sawlogs as shown above and the remaining pole.

Felling prescription

c. Trees felled in 10 metre strips.

d. Strips to be felled in the direction of extraction.

e. Converted produce *stacked in piles of 10 to 20 pieces,* at right-angles to and in the centre of the strip with pieces in the piles flush at the ends.

f. Logs and tops left in the felled strip but visible to the operator.

2 Job specification

The times apply to the following work:

Shortwood

Extraction of shortwood by one man on a Volvo SM 462 or SM 868 in the following stages:

a. Manoeuvre and travel inward to piles of produce.

b. Travel along lines of produce loading the forwarder with the optimum payload suitable for the travelling conditions, this includes double bunking products of 2 metres or less on the SM 462 and 3 metres or less on the SM 868.

c. Manoeuvre, travel back to the forest road and move on the road to off-load.

d. Off-load at right-angles and neatly stack produce in piles at roadside with ends flush.

Logs and tops
Extraction of logs and tops by one man on Volvo SM 462 Forwarder and SM 868 Grapple skidder.

a. Manoeuvre and travel inward to logs and/or tops.

b. Travel along lines of produce loading the forwarder or grapple skidder with the optimum payload for travelling conditions.

c. Manoeuvre, travel back to the forest road and move on the road to off-load.

d. With the forwarder or grapple skidder parallel to log pile and alongside, off-load and neatly stack logs and tops in piles at road-side, or unload in similar fashion at the customers premises as directed.

3 Tools and equipment (including safety equipment)

a. Volvo SM 868 with ÖSA 350 hydraulic crane for forwarding and ÖSA 820 inverted grapple for skidding or Volvo SM 462 with ÖSA 340P hydraulic crane for forwarding.

b. Tools and equipment for machine maintenance.

c. Protective clothing viz:
Safety helmet BS 2826 or 5240.
Boots with safety toecap.

4 Allowances

Included in the times

a. For work other than that of actual extraction (terminal/movement), eg move aside poles and brash, adjust load on Volvo, articulated trailer or ground, inspect and consider, filling drains with brash, debogging machine, repairs and maintenance of less than 15 minutes duration, prepare to load/unload, set up bolsters and pins on articulated trailers, tally loads etc.

  (i) *Volvo SM 462*
     Forwarding—21% of the time spent extracting.

(ii) *Volvo SM 868*
   Forwarding—21% of time spent extracting.
   Skidding—14% of time spent extracting.
   Loading articulated trailers—27% of time spent extracting.
b. For personal needs and rest 18–22% of the total working time.

5 Method of using the Output Guide

Paragraph 6A gives terminal times (loading, movement between piles and unloading) per load.

Paragraph 6B gives times for travelling.

a. Select the appropriate terminal time according to the product being handled, the mean payload being carried, and the product volume per hectare (SM 462, SABM pulp only).

b. Estimate mean distance travelled in the wood and read off the appropriate value for travelling.

c. Estimate the mean distance travelled along the forest road from the point of exit from the wood to where unloading takes place. Read off the appropriate value for travelling.

The sum of all these values will give the time per load in standard minutes. The output per day in loads can be obtained by dividing this figure into the total available minutes in the day.

Examples

1 Volvo SM 462, 2·5–3 metre SABM at 150 cubic metres/hectare, mean load 5 cubic metres, 500 metres travelling in the wood, 30 metres travelling along the forest road, unloaded to ground, 480 minutes per day.

	Paragraph 6	*Standard minutes*
Terminal time	A	30·3
(Paragraph 7A, deduct 3½% for 150 cubic metres per hectare)		$-1\cdot1$
		$\overline{29\cdot2}$
Travel in wood 500 metres	Bi	26·6
Travel on road 30 metres	Bii	1·0
Total cyclic time (time per load)	—	56·8
Total number of loads per day $=\dfrac{480}{56\cdot8}=8$		

2 Volvo SM 868, grapple skidding, sawlogs, 4 cubic metre mean load, 100 metres travelling in the wood, 150 metres travelling along the forest road, unloaded to ground, 480 minutes per day.

	Paragraph 6	Standard minutes
Terminal time	A	11·3
Travel in wood 100 metres	Bi	3·7
Travel on road 150 metres	Bii	3·1
Total cyclic time (time per load)	—	18·1
Total number of loads per day $=\dfrac{480}{18\cdot1}=26$		

6 Time in standard minutes per load for forwarding and skidding

A (i) *Terminal time (loading and unloading) for forwarding*
N.B. The standard minute values for SABM 2·5−3 metre pulp are based upon a product volume of 100 cubic metres per hectare. Modifications to this value for other volumes per hectare are given in paragraph 7A.

Product	Mean load size in cubic metres								
	SM 462					SM 868			
	4	5	6	7	8	9	10	11	12
2·5−3 m SABM	24·3	30·3	36·4	42·4	48·5	54·5	60·6	66·7	72·7
2·46 m ECC	23·0	28·7	34·4	40·2	45·9	51·7	57·4	63·1	68·9
1·85 m Woodcraft/ Coates 2 m Woodcemair 1·85 m ECC	29·6	37·0	44·4	51·8	59·2	66·6	74·0	81·4	88·8
3·66−8·23 m sawlogs	24·8	31·1	37·3	43·5	49·7	—	—	—	—
3·72−7·44 m sawlogs	24·1	30·1	36·1	42·1	48·2	54·2	60·2	66·2	72·2

When stakes are extracted by the SM 462 the terminal time is 16·4 standard minutes per cubic metre.

(ii) *Terminal time (loading and unloading) for skidding with the SM 868*

Product	Mean load size in cubic metres					
	2	3	4	5	6	7
0·37–3·66 m³ sawlog	5·6	8·5	11·3	14·2	17·0	19·8
0·11–0·5 m³ tops	9·2	13·8	18·4	23·1	27·6	32·2

B *Travelling time in the wood, on forest roads and on public roads in standard minutes per load*

(i) *Travelling in/out of wood (both ways)*

Distance (m)	SM 462 forwarding	SM 868 forwarding	SM 868 skidding
25	3·2	1·8	1·5
30	3·4	2·0	1·7
50	4·4	3·0	2·3
75	5·6	3·8	3·2
100	6·6	5·0	3·7
150	9·2	7·0	5·0
200	12·0	9·0	6·5
300	16·8	13·0	—
400	21·8	17·2	—
500	26·6	21·2	—
600	31·6	25·2	—

(ii) *Travelling on forest roads*

Distance (m)	SM 462 and SM 868 forwarding and skidding
10	0·4
20	0·7
30	1·0
50	1·5
75	2·0
100	2·5
150	3·1
200	3·7
300	4·6
400	5·6
500	6·6
600	7·5

6B(ii) *continued*

Distance (m)	SM 462 and SM 868 forwarding and skidding
800	9·3
1000	11·3
1200	13·2
1400	15·1
1600	17·0

(iii) *Travelling on public roads*

SM 462 and SM 868 7·7 standard minutes per mile (4·8 standard minutes per kilometre).

7 Modifications and variations to the time

N.B. These modifications and variations are only to be applied when the prevailing conditions or job specification differ from those listed in paragraphs 1 and 2.

A *Volvo SM 462 only. Terminal times*

The time values for 2·5–3 metre SABM pulp in paragraph 6A(i) are for a product volume of 100 cubic metres per hectare: when the volume per hectare varies adjust the time values as follows:

SABM pulp volume in cubic metres per hectare	Modification
50	Add 9%
150	Subtract $3\frac{1}{2}$%
200	Subtract 5%

B *Volvo SM 868 loading articulated trailers*

When unloading SABM material directly on to an articulated lorry trailer—

Add 3% to the terminal time (paragraph 6A).

C *Volvo SM 462 and SM 868 unloading sawlogs*

Where it is necessary for the forwarder or grapple skidder to unload and stack at right-angles to itself—

Add 15% to the terminal time (paragraph 6A).

D *Volvo SM 868 delivering logs to joiners yard*

Irrespective of the number of deliveries per day, add 12 standard minutes to the total cyclic times for reporting to mill office for instructions and for writing out and delivering conveyance notes.

Eg if three loads per day add four standard minutes per load.

Pole Length Ground Skidding by Roadless Hydrostatic Skidder

1 Conditions

The Output Guide applies to extraction under the following conditions:

a. Pole lengths.

b. First, line-thinning.

c. Tip first extraction.

d. Last whorl of branches left proud to prevent chokers from slipping off during extraction.

e. Poles felled directionally with tips pointing in the direction of extraction.

f. Tips of poles to be free of brash and points brought together wherever possible.

g. Tops of trees cut up to approximately 1·25m lengths.

h. Drains of 0·75m or more wide and 0·5m or more deep to be filled in with brash and be readily traversable.

i. Ample room for manoeuvring machine at roadside/conversion bay.

j. Slopes of up to 20% (11°).

2 Job specification

The Output Guide is for the following work:

a. Chokering carried out by tractor driver.

b. Unchokering carried out by convertor.

c. Tractor driven up rack, turned around and manoeuvred to face extraction direction prior to d. below.

d. Winch rope hauled out and chokers distributed to tip of each pole or poles to be extracted.

e. Poles chokered and secured to winch rope working back to tractor and winching in at the same time with the aid of the radio.

f. Repeat d. and c. above with the second winch rope, manoeuvring tractor if necessary.

g. Mount tractor and complete winching in load to butt plate with load lifted as high as possible from the ground.

h. Skid load to conversion point.

i. Drop load, convertors to pull out quick release pin and hang fresh set of chokers on butt plate pins.

j. Winch in rope and return to extract next load.

k. Carry out daily maintenance of tractor and equipment as instructed by supervisor.

3 Tools and equipment (including safety equipment)
a. Roadless hydrostatic skidder equipped with:
Radio controlled double drum winch.
Remote control radio pack and spare battery.
Soft loop at end of each winch rope.
Two sets of detachable choker hooks with polypropylene rope chokers (at least 30 chokers and hooks).
Two quick release pins (plus one spare).
Appropriate maintenance tools.

b. Protective clothing viz:
Safety helmet BS 2826 or 5240.
Ear defenders.
Gloves with wireproof palms.
Boots with safety toecap.

4 Allowances
Included in the Output Guide:

a. For work other than actual extraction—18% of terminal and movement time.

b. For personal needs and rest—20% of the total working time.

5 Method of using the Output Guide
Entry to paragraph 6 is gained by using average volume of poles to be extracted and average extraction distance. (An indication of expected load size is given and output is controlled by this average load size.)

6A   Time in standard minutes per m³

Pole size in m³	Number of poles	Load size in m³	Extraction distance in metres				
			100	200	300	400	500
0·02	26	0·52	32·96	39·58	46·19	52·81	59·42
0·04	23	0·92	18·63	22·37	26·11	29·85	33·59
0·06	20	1·20	14·28	17·15	20·02	22·88	25·75
0·08	16	1·28	13·39	16·08	18·77	21·45	24·14
0·10	13	1·30	13·19	15·83	18·48	21·12	23·77
0·12	11	1·32	12·98	15·59	18·20	20·80	23·41
0·14	10	1·40	12·24	14·70	17·16	19·61	22·07

6B   Time in standard minutes per pole

Pole size in m³	Number of poles	Load size in m³	Extraction distance in metres				
			100	200	300	400	500
0·02	26	0·52	0·66	0·79	0·92	1·06	1·19
0·04	23	0·92	0·75	0·90	1·04	1·19	1·34
0·06	20	1·20	0·86	1·03	1·20	1·37	1·55
0·08	16	1·28	1·07	1·29	1·50	1·72	1·93
0·10	13	1·30	1·32	1·58	1·85	2·11	2·38
0·12	11	1·32	1·56	1·87	2·18	2·50	2·81
0·14	10	1·40	1·71	2·06	2·40	2·75	3·09

7 Modifications
None.

Extraction by Ploughmaster '46' Fitted with Igland 4 Tonne Double Drum Winch
Borders

1 **Conditions**

The Output Guide applies to the extraction of tree lengths or trees crosscut at stump into sawlog/s and the remaining pole (see paragraph 5a) under the following conditions:

a. Downhill extraction.

b. Undulating terrain, average slope 15%, maximum 33% (8·5°–18°).

c. Windblown areas with no formal presentation for extraction or clearfelled areas on which trees have been felled out from a working face suitable for tip first extraction (for thinnings see paragraph 7a).

d. Soil: Peat over clay or shaley ground. Two types of ground conditions are identified:
(i) **Soft** where the peat averages up to 23 cm in depth.
(ii) **Firm** where the average depth of peat does not exceed 8 cm.

e. Drains and ditches in the extraction routes to be bridged or filled in with brash; stumps to be cut as low as practicable.

2 **Job specification**

The Output Guide is for the following work:

a. Chokering and unchokering the load together with the actual extraction (see paragraph 7b where the actual unchokering is carried out by crosscutters).

b. Trees from clearfelling areas to be extracted tip first, 'pegs' having been left on the last 2 whorls to help retain the chokers.

c. Tree lengths and the tops of poles with the sawlogs removed to be stacked at roadside as instructed by the supervisor.

d. Sawlogs to be stacked separately at roadside as instructed by the supervisor.

3 **Tools and equipment** (including safety equipment)

a. Ploughmaster '46' tractor fitted with approved safety cab; Igland 'Spesial' 4 tonne, 2 drum winch, wheel chains, valve guards, radiator guard and adequate underbody protection.

b. Eight chain chokers with sliding hooks and keyhole adjustors—4 to each rope.

c. Tools for maintenance and adjustment of tractor and equipment.

d. Quick release snatch block with sling or anchor hook.

e. Protective clothing viz:
Safety helmet BS 2826 or 5240.
Ear defenders.
Gloves with wireproof palms.
Boots with safety toecap.

4 Allowances

The following allowances are included in the Output Guide:

a. For contingencies and work other than actual extraction (terminal and movement), eg refuelling tractor (where fuel supply is adjacent to the working area), routine maintenance of the tractor, adjusting wheel chains, adjusting winch, repairing winch rope, clearing brash, a minimal amount of route preparation, inspecting and considering, minor de-bogging and snagging; but **not** clearance of drains after extraction: 18% of the time spent extracting.

b. For personal needs and rest, 25% of the terminal time, 19% of the movement time.

5 Method of using the Output Guide

a. This Output Guide caters for two methods of extraction from either windblow or clearfelling (see paragraph 7a for guidance on thinnings). Produce is extracted either as a tree length or as a sawlog/s and remaining pole after primary conversion has taken place at stump.

Tables 6A(i), (ii), (iii) and (iv) give terminal time (loading and unloading) in standard minutes per load and table 6B gives movement time (in and out) in standard minutes per load against the mean extraction distance (out only).

b. Mean extraction distance.

The mean extraction distance is assessed either by measurements from a map or by actual measurements taken on the ground of
(i) Racks or extraction routes.
(ii) From rack outlet/s to conversion/stacking site/s. The sum of these two mean distances is the mean extraction distance.

c. Selection of the time.

The following information is required before the time can be selected:
(i) The average volume of felled or blown trees (see FC Booklet No 36).

344

(ii) The mean extraction distance to off-loading point.
(If the tables do not extend to the total distance required, the time for movement under **soft** conditions is 0·08 standard minutes per metre and under **firm** conditions 0·05 standard minutes per metre and the time for movement can be calculated by multiplying the extraction distance by the appropriate time per metre.)

(iii) Ground conditions, ie whether to be categorised as **soft** or **firm** (see paragraph 1d).

(iv) Whether tree lengths or sawlog/s and remaining pole.

(v) Clearfelling or windblow.

N.B. The time for extraction is terminal time (6A (i), (ii), (iii) or (iv)) plus movement time (6B).

Example
Clearfelling
Average volume per tree 0·14 m³
Tree length extraction

Total rack length 300 m $\therefore$ mean rack length $\dfrac{300}{2} = 150$ m

Mean extraction distance from rack outlet to conversion site = 25 m
Total mean extraction distance = 175 m
Average depth of peat = 5 cm
Chokers as shown in paragraph 3b
Movement time =   8·75 standard minutes per load
                (from 6B firm conditions)
Terminal time   = 10·10 standard minutes per load
                (from 6A(iii))
Total time   = 18·85 standard minutes per load

If the time is required as a time per tree, calculate the number of trees per load by dividing the average load size by the average volume per tree
eg 1·10÷0·14 = 7·86 trees per load
and then the time per load by the average number of trees per load
eg 18·85÷7·86 = 2·40 standard minutes per tree
If the time is required as a time per m³ use the following equation:

$$\frac{1}{\text{Average load size in m}^3} \times \text{time per load} = \text{time per m}^3$$

eg $\dfrac{1}{1·10} \times 18·85 = 17·14$ standard minutes per m³

Example
Conditions as above except that detachable choker hooks and rope chokers are used and chokers are removed by crosscutters.
Movement time =   8·75 standard minutes per load
                (from 6B firm conditions)

Terminal time = 10·10 standard minutes per load
(from 6A(iii) reduced by 18% (from 7b))
= 10·10 − 1·82
= 8·28 standard minutes per load

Total time = 8·75 + 8·28 = 17·03 standard minutes per load

6A   Terminal times

(i)   *Windblow—tree lengths. Time per load in standard minutes.*

Average tree size (m³)	0·07	0·11	0·14	0·18	0·22	0·25–0·36
Associated average load size (m³)	0·70	0·70	1·10	1·10	1·10	1·50
Standard minutes per load	14·9	12·6	16·8	14·6	13·1	16·6

(ii)   *Windblow—sawlogs and remaining pole. Time per load in standard minutes.*

Average tree size (m³)	0·07	0·11	0·14	0·18	0·22	0·25–0·36
Associated average load size (m³)	0·70	0·70	1·10	1·10	1·10	1·50
Standard minutes per load	18·6	15·3	19·0	16·3	14·5	17·8

(iii)   *Clearfelling—tree lengths. Time per load in standard minutes.*

Average tree size (m³)	0·07	0·11	0·14	0·18	0·22	0·25–0·36
Associated average load size (m³)	0·70	0·70	1·10	1·10	1·10	1·50
Standard minutes per load	9·3	7·6	10·1	8·8	8·2	10·8

(iv) *Clearfelling—sawlogs and remaining pole. Time per load in standard minutes.*

Average tree size (m³)	0·07	0·11	0·14	0·18	0·22	0·25–0·36
Associated average load size (m³)	0·70	0·70	1·10	1·10	1·10	1·50
Standard minutes per load	13·0	10·3	12·3	10·5	9·6	12·0

6B Movement times

Time per load in standard minutes (average load size between 0·70 m³ and 1·50 m³).

Mean extraction distance in m (see paragraph 5)	Soft conditions (see paragraph 1d) Time per load in standard minutes	Firm conditions (see paragraph 1d) Time per load in standard minutes
50	4·00	2·50
75	6·00	3·75
100	8·00	5·00
125	10·00	6·25
150	12·00	7·50
175	14·00	8·75
200	16·00	10·00
225	18·00	11·25
250	20·00	12·50
275	22·00	13·75
300	24·00	15·00

7 Modifications to the Output Guide

N.B. These modifications and variations are only to be applied when the prevailing conditions or job specification differ from those listed in paragraphs 1 and 2.

a. *Thinning*

The following *tentative* guidance is given.

Terminal time similar to that for windblow (paragraphs 6A(i) or 6A(ii)).

Travelling time similar to paragraph 6B.

b. *Detachable choker hooks and rope chokers*

If the driver is using detachable choker hooks and rope chokers and is not unchokering but simply unloading by removing the load retaining pin and then leaving chokered trees and collecting a fresh supply of hooks and chokers for his next load the terminal times can be reduced by up to 18% (the chokers will be removed from the load by the crosscutters).

	Section	XXII
Forestry Commission Cable Crane (Skyline) Sawlogs and Pulpwood	No	G11

Forestry Commission Cable Crane (Skyline) Sawlogs and Pulpwood
North and West Scotland and North West England

1 **Conditions**

The Output Guide applies to extraction under the following conditions:

a. Pulpwood cut to either Scottish pulpwood or TBM pulpwood specifications.
(Scottish pulp: Length 3·0 m with tolerance down to 2·5 m; diameter 5–40 cm over bark.
Thames Board Mill: Length 2·3 m with tolerance down to 1·8 m; diameter 6·5–40 cm over bark.)

b. Sawlogs cut to local specifications.

c. Loads of pulpwood to have one end raised on a bearer or be positioned by the feller so that they can be easily chokered.

d. Loads to be positioned by the feller so that they can be pulled to the Skyline with the minimum of obstruction.

e. Racks in thinnings to be at least 3 m wide, straight, and spaced at intervals of 20 m–30 m. In clearfelling and windblow, strips 10 m–12 m wide to be worked.

f. Slopes up to 36% (20°) (see paragraph 7c for steeper slopes).

g. Few boulders but normal quantities of brash and tops.

h. Loads extracted either uphill or downhill.

i. Tractor either in line with, or offset, to the rack (but see paragraph 7i).

j. The chokerman able to catch the chain for 'haul-back no brakes' method of rope pay-off (see paragraph 7d for situations where this is not possible).

k. A team of two men.

2 **Job specification**

a. All produce to be extracted and stacked at roadside (see paragraph 7a when the stack height is more than 3 m or spreads over more than 7 m on either side of the Skyline).

b. Stacks to have all ends flush together suitable for subsequent mechanical loading on to lorries. Stacks can be either at right angles, or parallel, to the road.

c. Individual products to be stacked separately.

3 Tools and equipment (including safety equipment)

a. Igland Spesial 4000 double drum winch modified and equipped by Chapelhall workshops to mechanical plant specification 75/57. Mounted on a Ford 3000 Selectospeed, Ford 3000, British Leyland 245 or Massey Ferguson 135 tractor.

N.B. These tractors give different outputs—see paragraph 6.

b. Two radio sets. (One pack set with a mobile and loudspeaker at the tractor.)

c. Whip aerials to be available when reception difficulties occur.

d. EIA or Parkgate hooks with polypropylene rope chokers.

e. Protective clothing viz:
Safety helmet BS 2826 or 5240.
Ear defenders.
Gloves with wireproof palms.
Boots with safety toecap.
Chokerman need not wear ear defenders.
In addition, outer clothing must be devoid of any loose flaps likely to get caught in the moving parts of the machinery or running cables.

4 Allowances

Included in the Output Guide:

a. For contingencies and work other than extraction, eg refuelling, daily maintenance, to and from camp, moving tractor to let road vehicles pass, counting and checking, improving stacking sites and running repairs 20% of the time spent on extraction.

N.B. Time for long splicing ropes or other jobs of more than 15 minutes duration are not included.

b. For personal needs and rest, 26% of the total working time.

5 Method of using the Output Guide

a. The Output Guide is composed of separate tables for set-up and take-down (6A), and for extraction (6B, C and D). The following facts are required to select the time for each part:

(i) The average length of rack measured from roadside to the foot of the spar tree.

(ii) The total volume coming from the area to be extracted.

(iii) The average load size (see paragraph 5c(ii) for details).

(iv) The type of tractor in use (ie Ford 3000 Selectomatic or British Leyland 245, Ford 3000, Massey Ferguson 135).

(v) Thinning or clearfelling/windblow.

(vi) Product.

b. *Set-up and take-down*

The time per cubic metre for set-up and take-down is dependent on rack espacement (or number of racks in the area to be extracted), volume yield of the area and rack length. The steps to be taken to determine the time per cubic metre are:

(i) Divide the total volume to be removed by the number of racks in the area. (Gives volume per rack.)

(ii) From table 6A read off the time per rack for set-up and take-down against the appropriate rack length.

(iii) Divide the time per rack (from (ii) above) by the volume per rack. (Gives time per m^3 for set-up and take-down.)

c. *Extraction*

The time per cubic metre for extraction is dependent on rack length, produce type, average load size, thinning or clearfelling/windblow and tractor type. The steps to be taken to determine the time per cubic metre are:

(i) Determine the average rack length. (The average length of rack can be measured by pacing on the ground the length of a sample of racks (eg every third). On broken, hillocky ground, rack length will be very variable and probably every rack will have to be measured. The easiest time to note the rack length is at the time of marking.)

(ii) Determine the average load size and volume to be extracted for each product type. This is best assessed by going over the area after felling is complete and by measuring sample loads. Multiple chokers should be used wherever possible to increase the average load size (within the limits shown in paragraph 6).

(iii) The time per m^3 is then read off from paragraph 6 against rack length and load volume for the appropriate tractor against either clearfelling/windblow or thinning and either pulpwood or sawlogs.

(iv) If an overall time per m^3, regardless of product type, is required then the time for each product type must be assessed and then weighted according to the product mix, there are two ways of doing this, eg

45 m^3 per ha to be extracted from thinnings

15 m^3 per ha of sawlogs averaging 0·30 m^3 per load

30 m^3 per ha of pulpwood averaging 0·35 m^3 per load

Rack length 300 m

Ford 3000 Selectospeed.

Product	Time in standard minutes per m³ for product	Volume to be extracted	% Product type	Time per m³ for product mix $\frac{col\ 2 \times 4}{100}$
Sawlogs	36·29 (from 6B(iv))	15	33·33	12·09
Pulpwood	45·37 (from 6B(iii))	30	66·67	30·25
Total		45	100	42·34 SMs

or

Product	Time in standard minutes per m³ for product	Volume to be extracted	Total time for product col 2 × 3
Sawlogs	36·29	15	544
Pulpwood	45·37	30	1361
Total		45	1905

The weighted time per m³ is then calculated $\dfrac{1905}{45} = 42 \cdot 34$ SMs

N.B. **The times shown in paragraph 6 (and the modifications in paragraph 7) apply to a two-man team.**

For use in forecasting machine requirements the times should be halved, eg time for two men to set-up and take-down is 330 standard minutes.

Machine time $= \dfrac{330}{2} = 165$ SMs.

Where output is required in terms of m³ per hour the following calculation should be carried out.

$60 \div \left(\dfrac{\text{Time per m}^3 \text{ in SMs}}{2} \right) =$ output per machine hour

eg Time per m³ for extraction $= 43 \cdot 76$ SMs

Output per machine hour $= 60 \div \left(\dfrac{43 \cdot 76}{2} \right) = 2 \cdot 74$ m³.

Example
Average rack length = 180 m
Thinning
Racks 25 m apart
20 m³ extracted per rack (Total)
15 m³ Scottish pulpwood per rack average load size 0·30 m³
 5 m³ sawlogs per rack   average load size 0·20 m³
Slope 20% (11°)
Offset head spar
Ford 3000 Selectospeed tractor.

Set-up and take-down
Time for set-up and take-down = 310 SMs (from 6A)
+ 30 SMs for offset head spar
(from 7i)
= 340 SMs.

Time per m³ for set-up and take-down $= \dfrac{340}{20 \, \text{m}^3 \text{ per rack}}$
= 1700 SMs.

Extraction
Time per m³ for Scottish pulpwood = 41·10 SMs
Time per m³ for sawlogs   = 50·45 SMs

Weighted time per m³ $= \dfrac{(41{\cdot}10 \times 15) + (50{\cdot}45 \times 5)}{20}$ SMs

= 43·44 SMs.

Total time per m³
The time in standard minutes per m³ for extracting 1 rack
= 17·00 (set-up and take-down)
+ 43·44 (extraction)
= 60·44 SMs.

Machine requirement

Machine time for 1 rack $= \dfrac{60{\cdot}44 \, \text{SMs} \times 20 \, \text{m}^3}{2}$

= 604·4 SMs.

'Calculated' piecework rate
Price per standard minute = 1·99p
Price per m³   = 1·99p × 60·44 SMs
= £1·20.

N.B. As the times are for two men the price per standard minute
for one man is used to arrive at the calculated rate.

6A   Time in standard minutes per rack for set-up and take-down
Thinning and clearfelling
N.B. These times are for both uphill and downhill working. See
paragraph 7 for additions to be made when supports, ground anchors
or offset working are employed. A deduction is required if the tail
spar is not moved. (See paragraph 7.)

Rack length	Time in standard minutes per rack for set-up and take-down
120	248
140	269
160	289
180	310
200	330

Rack length	Time in standard minutes per rack for set-up and take-down
220	351
240	372
260	392
280	413
300	434
320	454
340	475
360	496
380	516
400	537

N.B. These times are for a two-man team.

6B(i) Time in standard minutes per m³ for extraction by Forestry Commission cable crane (Skyline) mounted on a Ford 3000 Selectospeed or British Leyland 245 (with 2-speed pto) tractor

Clearfelling and windblow only—pulpwood

Rack length m	Standard minutes per m³ for extraction					
	Average load size m³					
	0·50	0·45	0·40	0·35	0·30	0·25
120	24·57	25·97	27·72	29·97	32·97	37·17
140	25·70	27·22	29·13	31·58	34·85	39·42
160	26·82	28·46	30·53	33·18	36·71	41·66
180	27·94	29·71	31·93	34·78	38·59	43·91
200	29·06	30·96	33·33	36·38	40·45	46·15
220	30·18	32·21	34·74	37·99	42·33	48·40
240	31·30	33·45	36·14	39·59	44·19	50·63
260	32·43	34·70	37·54	41·19	46·07	52·88
280	33·55	35·94	38·94	42·79	47·93	55·12
300	34·67	37·19	40·35	44·40	49·81	57·37
320	35·79	38·44	41·75	46·00	57·67	59·61
340	36·92	39·68	43·15	47·60	53·55	61·86
360	38·04	40·93	44·55	49·20	55·41	64·10
380	39·16	42·18	45·96	50·81	57·29	66·35
400	40·28	43·42	47·36	52·41	59·15	68·58

N.B. The times are for a team of two men.

6B(ii) **Time in standard minutes per m³ for extraction by Forestry Commission cable crane (Skyline) mounted on a Ford 3000 Selectospeed or British Leyland 245 (with 2-speed pto) tractor**

Clearfelling and windblow only—sawlogs

Rack length m	Standard minutes per m³ for extraction					
	Average load size m³					
	0·50	0·45	0·40	0·35	0·30	0·25
120	25·06	26·28	27·81	29·77	32·40	36·07
140	25·88	27·19	28·83	30·95	33·77	37·72
160	26·70	28·10	29·86	32·12	35·13	39·35
180	27·52	29·01	30·89	33·29	36·50	40·99
200	28·34	29·92	31·91	34·46	37·87	42·64
220	29·16	30·83	32·94	35·64	39·24	44·27
240	29·98	31·75	33·97	36·81	40·60	45·92
260	30·80	32·66	34·99	37·98	41·98	47·56
280	31·62	33·57	36·02	39·15	43·35	49·19
300	32·44	34·48	37·04	40·33	44·71	50·84
320	33·26	35·40	38·07	41·50	46·08	52·48
340	34·09	36·31	39·09	42·67	47·45	54·13
360	34·90	37·22	40·12	43·85	48·81	55·76
380	35·73	38·13	41·15	45·02	50·18	57·41
400	36·55	39·04	42·17	46·19	51·55	59·05

N.B. The times are for a team of two men.

6B(iii) **Time in standard minutes per m³ for extraction by Forestry Commission cable crane (Skyline) mounted on a Ford 3000 Selectospeed or British Leyland 245 (with 2-speed pto) tractor**

Selective thinning only—pulpwood

Rack length m	Standard minutes per m³ for extraction			
	Average load size m³			
	0·40	0·35	0·30	0·25
120	30·65	33·09	36·33	40·84
140	31·85	34·46	37·90	42·76
160	33·05	35·80	39·50	44·68
180	34·25	37·17	41·10	46·60
200	35·43	38·54	42·70	48·48
220	36·63	39·91	44·27	50·40
240	37·83	41·26	45·87	52·32
260	39·03	42·63	47·47	54·24

Rack length m	Standard minutes per m³ for extraction			
	Average load size m³			
	0·40	*0·35*	*0·30*	*0·25*
280	40·20	44·00	49·07	56·16
300	41·40	45·37	50·67	58·04
320	42·60	46·74	52·23	59·96
340	43·80	48·09	53·83	61·84
360	45·00	49·46	55·43	63·88
380	46·18	50·83	57·03	65·68
400	47·38	52·20	58·63	67·60

N.B. The times are for a team of two men.

6B(iv) **Time in standard minutes per m³ for extraction by Forestry Commission cable crane (Skyline) mounted on a Ford 3000 Selectospeed or British Leyland 245 (with 2-speed pto) tractor**
Selective thinning only—sawlogs

Rack length m	Standard minutes per m³ for extraction					
	Average load size m³					
	0·40	*0·35*	*0·30*	*0·25*	*0·20*	*0·15*
120	21·95	25·09	29·27	35·12	43·90	58·53
140	23·05	26·34	30·73	36·88	46·10	61·47
160	24·13	27·57	32·17	38·60	48·25	64·33
180	25·23	28·83	33·63	40·36	50·45	67·27
200	26·30	30·06	35·07	42·08	52·60	70·13
220	27·40	31·31	36·53	43·84	54·80	73·07
240	28·48	32·54	37·97	45·56	56·95	76·93
260	29·58	33·80	39·43	47·32	59·15	78·87
280	30·68	35·06	40·90	49·08	61·35	81·80
300	31·75	36·29	42·33	50·80	63·50	84·67
320	32·85	37·54	43·80	52·56	65·70	87·60
340	33·93	38·77	45·90	54·28	67·85	90·47
360	35·03	40·03	46·70	56·04	70·05	93·40
380	36·10	41·26	48·13	57·76	72·20	96·27
400	37·20	42·51	49·60	59·52	74·40	99·20

N.B. The times are for a team of two men.

6C(i) Time in standard minutes per m³ for extraction by Forestry Commission cable crane (Skyline) mounted on a Ford 3000 tractor

Clearfelling and windblow only—pulpwood

Rack length m	Standard minutes per m³ for extraction					
	Average load size m³					
	0·50	0·45	0·40	0·35	0·30	0·25
120	24·08	25·42	27·10	29·26	32·15	36·18
140	25·40	26·89	28·76	31·15	34·35	38·83
160	26·72	28·36	30·41	33·04	36·55	41·47
180	28·04	29·83	32·06	34·93	38·76	44·11
200	29·36	31·29	33·71	36·81	40·96	46·75
220	30·69	32·76	35·37	38·70	43·16	49·40
240	32·01	34·23	37·01	40·59	45·36	52·04
260	33·33	35·70	38·67	42·48	47·57	54·69
280	34·65	37·17	40·32	44·36	49·77	57·32
300	35·97	38·64	41·97	46·26	51·97	59·97
320	37·29	40·10	43·62	48·14	54·17	62·61
340	38·62	41·57	45·28	50·03	56·38	65·26
360	39·94	43·04	46·93	51·92	58·58	67·90
380	41·26	44·51	48·58	53·81	60·78	70·54
400	42·58	45·97	50·23	55·69	62·98	73·18

N.B. The times are for a team of two men.

6C(ii) Time in standard minutes per m³ for extraction by Forestry Commission cable crane (Skyline) mounted on a Ford 3000 tractor

Clearfelling and windblow only—sawlogs

Rack length m	Standard minutes per m³ for extraction					
	Average load size m³					
	0·50	0·45	0·40	0·35	0·30	0·25
120	24·56	25·73	27·19	29·06	31·57	35·08
140	25·58	26·86	28·46	30·52	33·27	37·12
160	26·60	27·99	29·74	31·98	34·97	39·16
180	27·62	29·13	31·02	33·44	36·67	41·20
200	28·64	30·26	32·29	34·89	38·38	43·24
220	29·66	31·39	33·57	36·35	40·08	45·28
240	30·68	32·53	34·84	37·81	41·77	47·32
260	31·70	33·67	36·11	39·27	43·48	49·36

6C(ii)—*continued*

Rack length m	Standard minutes per m³ for extraction					
	Average load size m³					
	0·50	*0·45*	*0·40*	*0·35*	*0·30*	*0·25*
280	32·72	34·79	37·39	40·73	45·18	51·40
300	33·74	35·93	38·66	42·19	46·87	53·44
320	34·76	37·06	39·94	43·64	48·58	55·48
340	35·79	38·19	41·22	45·10	50·28	57·53
360	36·81	39·33	42·50	46·56	51·98	59·56
380	37·82	40·46	43·77	48·01	53·68	61·61
400	38·85	41·59	45·04	49·47	55·38	63·65

N.B. The times are for a team of two men.

6C(iii) **Time in standard minutes per m³ for extraction by Forestry Commission cable crane (Skyline) mounted on a Ford 3000 tractor**

Selective thinning only—pulpwood

Rack length m	Standard minutes per m³ for extraction			
	Average load size m³			
	0·40	*0·35*	*0·30*	*0·25*
120	30·03	32·37	35·50	39·84
140	31·48	34·03	37·43	42·16
160	32·93	35·69	39·33	44·48
180	34·38	37·31	41·27	46·80
200	35·80	38·97	43·20	49·08
220	37·25	40·63	45·13	51·40
240	38·70	42·29	47·03	53·72
260	40·15	43·91	48·97	56·04
280	41·58	45·57	50·90	58·36
300	43·03	47·23	52·83	60·72
320	44·48	48·89	54·73	62·96
340	45·93	50·51	56·67	65·28
360	47·35	52·17	58·60	67·60
380	48·80	53·83	60·53	69·88
400	50·25	55·49	62·43	72·20

N.B. The times are for a team of two men.

6C(iv) Time in standard minutes per m³ for extraction by Forestry Commission cable crane (Skyline) mounted on a Ford 3000 tractor

Selective thinning only—sawlogs

Rack length m	Standard minutes per m³ for extraction					
	Average load size m³					
	0·40	*0·35*	*0·30*	*0·25*	*0·20*	*0·15*
120	21·33	24·37	28·43	34·12	42·65	56·87
140	22·68	25·91	30·23	36·28	45·35	60·47
160	24·00	27·43	32·00	38·40	48·00	64·00
180	25·35	28·97	33·80	40·56	50·70	67·60
200	26·68	30·49	35·57	42·68	53·35	71·13
220	28·03	32·03	37·37	44·84	56·05	74·73
240	29·38	33·57	39·17	47·00	58·75	78·33
260	30·70	35·09	40·93	49·12	61·40	81·87
280	32·05	36·63	42·73	51·28	64·10	85·47
300	33·38	38·14	44·50	53·40	66·75	89·00
320	34·73	39·69	46·30	55·56	69·45	92·60
340	36·05	41·20	48·07	57·68	72·10	96·13
360	37·40	42·74	49·87	59·84	74·80	99·73
380	38·73	44·26	51·63	61·96	77·45	103·27
400	40·08	45·80	53·43	64·12	80·15	106·87

N.B. The times are for a team of two men.

6D(i) Time in standard minutes per m³ for extraction by Forestry Commission cable crane (Skyline) mounted on a Massey Ferguson 135 tractor

Clearfelling and windblow only—pulpwood

Rack length m	Standard minutes per m³ for extraction					
	Average load size m³					
	0·50	*0·45*	*0·40*	*0·35*	*0·30*	*0·25*
120	24·60	26·00	27·75	30·01	33·01	37·22
140	25·81	27·34	29·26	31·73	35·03	39·64
160	27·01	28·68	30·77	33·45	37·04	42·05
180	28·22	30·02	32·28	35·18	39·05	44·46
200	29·42	31·36	33·79	36·90	41·06	46·87
220	30·63	32·70	35·30	38·63	43·07	49·29
240	31·84	34·04	36·80	40·35	45·08	51·70
260	33·05	35·38	38·31	42·07	47·10	54·12

6D(i)—*continued*

Rack length m	*Standard minutes per m³ for extraction*					
	Average load size m³					
	0·40	*0·35*	*0·30*	*0·25*	*0·20*	*0·15*
280	34·25	36·72	39·82	43·79	49·10	56·53
300	35·46	38·06	41·33	45·52	51·12	58·94
320	36·66	39·40	42·84	47·24	53·12	61·35
340	37·87	40·75	44·35	48·97	55·14	63·77
360	39·08	42·08	45·85	50·69	57·15	66·18
380	40·29	43·43	47·36	52·42	59·16	68·60
400	41·49	44·76	48·87	54·14	61·17	71·00

N.B. The times are for a team of two men.

6D(ii) **Time in standard minutes per m³ for extraction by Forestry Commission cable crane (Skyline) mounted on a Massey Ferguson 135 tractor**

Clearfelling and windblow only—sawlogs

Rack length m	*Standard minutes per m³ for extraction*					
	Average load size m³					
	0·50	*0·45*	*0·40*	*0·35*	*0·30*	*0·25*
120	25·08	26·31	27·84	29·91	32·44	36·12
140	25·99	27·31	28·97	31·10	33·95	37·93
160	26·89	28·31	30·10	32·40	35·45	39·74
180	27·80	29·32	31·24	33·69	36·96	41·55
200	28·70	30·32	32·36	34·98	38·48	43·36
220	29·61	31·33	33·50	36·28	39·99	45·17
240	30·51	32·34	34·63	37·57	41·49	46·98
260	31·42	33·35	35·76	38·86	43·00	48·79
280	32·32	34·35	36·89	40·16	44·52	50·60
300	33·23	35·36	38·03	41·45	46·02	52·41
320	34·14	36·37	39·15	42·74	47·53	54·23
340	35·04	37·37	40·29	44·04	49·04	56·04
360	35·94	38·37	41·42	45·33	50·54	57·84
380	36·85	39·38	42·56	46·62	52·06	59·66
400	37·76	40·38	43·68	47·92	53·57	61·47

N.B. The times are for a team of two men.

6D(iii) **Time in standard minutes per m³ for extraction by Forestry Commission cable crane (Skyline) mounted on a Massey Ferguson 135 tractor**

Selective thinning only—pulpwood

Rack length m	Standard minutes per m³ for extraction			
	Average load size m³			
	0·40	0·35	0·30	0·25
120	30·68	33·11	36·37	40·92
140	31·98	34·60	38·10	43·00
160	33·28	36·09	39·83	45·08
180	34·58	37·57	41·57	47·16
200	35·88	39·06	43·30	49·24
220	37·18	40·54	45·03	51·32
240	38·48	42·03	46·77	53·40
260	39·78	43·51	48·50	55·48
280	41·08	45·00	50·23	57·56
300	42·38	46·49	51·97	59·64
320	43·70	47·97	53·70	61·72
340	45·00	49·46	55·43	63·80
360	46·30	50·94	57·17	65·88
380	47·60	52·43	58·90	67·96
400	48·90	53·91	60·63	70·04

N.B. The times are for a team of two men.

6D(iv) **Time in standard minutes per m³ for extraction by Forestry Commission cable crane (Skyline) mounted on a Massey Ferguson 135 tractor**

Selective thinning only—sawlogs

Rack length m	Standard minutes per m³ for extraction					
	Average load size m³					
	0·40	0·35	0·30	0·25	0·20	0·15
120	21·98	25·11	29·30	35·16	43·95	58·60
140	23·18	26·49	30·90	37·08	46·35	61·80
160	24·38	27·86	32·50	39·00	48·75	65·00
180	25·58	29·23	34·10	40·92	51·15	68·20
200	26·78	30·60	35·70	42·84	53·55	71·40
220	27·95	31·94	37·27	44·72	55·95	74·53
240	29·15	33·31	38·87	46·64	58·30	77·73
260	30·35	34·69	40·47	48·56	60·70	80·93

Rack length m	Standard minutes per m³ for extraction					
	Average load size m³					
	0·50	*0·45*	*0·40*	*0·35*	*0·30*	*0·25*
280	31·55	36·06	42·07	50·48	63·10	84·13
300	32·73	37·40	43·63	52·36	65·45	87·27
320	33·93	38·77	45·23	54·28	67·85	90·47
340	35·13	40·14	46·83	56·20	70·25	93·67
360	36·33	41·51	48·43	58·12	72·65	98·87
380	37·53	42·89	50·03	60·04	75·05	100·07
400	38·70	44·23	51·60	61·92	77·40	103·20

N.B. The times are for a team of two men.

7 Modifications and variations to the Output Guide

N.B. These modifications and variations are only to be applied when the prevailing conditions or job specification differ from those listed in paragraphs 1 and 2.

a. *Pulpwood—difficult stacking conditions*
  (i) Where the stack height is over 3 m
      Add 5% to the extraction times.
  (ii) Where the stack spreads over more than 7 m on each side of the winch
      Add 5% to the extraction times.

These additions must be made to every load extracted in racks where they are justified. The conditions should be assessed before the job begins but due allowance must be made for possible interim lorry uplifting.

b. *Bad ground conditions*
  Where the ground is boulder strewn, gullied and heavily brash covered
      Add up to 5% in $2\frac{1}{2}$% steps.

c. *Slope*
  (i) Where the slope is over 36% (20°)
      Add $2\frac{1}{2}$%.
  (ii) Where the slope is over 58% (30°)
      Add 5%.

d. *Stop plate line pay-off*
  If used instead of 'haul back no brakes' method on a very concave profile
      Add 0·80 SMs per load to the extraction times.

e. *Set-up and take-down on M-type support*
   To set-up and take-down on an M-type support
   Add 36 SMs to the time for set-up and take-down.

f. *Set-up and take-down on A-frame support*
   To set-up and take-down on an A-frame support
   Add 67 SMs to the time for set-up and take-down.

g. *Difficult anchoring conditions*
   Where anchoring conditions are difficult
   Add up to 12 SMs in two steps to the time for set-up and take-down.

h. *Ground anchors*
   For each twin plate set of ground anchors used in set-up and take-down
   Add 20 SMs to the time for set-up and take-down.

i. *Offset head spar*
   To rig and de-rig an offset head spar
   Add 30 SMs to the time for set-up and take-down.

j. *Tail spar not moved*
   If the tail spar is not moved
   Deduct 60 SMs from the time for set-up and take-down.

k. *Tractor engine speed limited to 1800 rpm*
   Where the tractor engine speed is limited to a maximum of 1800 rpm the times for extraction must be *increased* by the percentages shown in the table below.

Rack length m	Tractor type		
	Ford 3000 Selectospeed and BL 245	Ford 3000	MF 135
120	4·26	3·95	3·60
140	4·40	4·17	3·83
160	4·55	4·40	4·06
180	4·69	4·62	4·29
200	4·84	4·85	4·52
220	4·98	5·07	4·75
240	5·13	5·30	4·98
260	5·27	5·54	5·21
280	5·42	5·76	5·44
300	5·56	5·99	5·67
320	5·71	6·21	5·90
340	5·85	6·44	6·13
360	6·00	6·66	6·36
380	6·14	6·89	6·59
400	6·29	7·12	6·82

Loading, Carriage and Unloading Lorries Equipped with a 2 Tonne Hiab Crane and Grapple

East England

1 Conditions

The Output Guide applies to loading, carriage and unloading lorries under the following conditions:

a. Crane-equipped lorry and produce categories:

Product specification		Nominal carrying capacity of vehicle	Product type	For terminal times see paragraph 6A, table number
Length (m)	Diameter UB (cm)			
4–8	7–20	12 tonnes	Cambio poles	(i)
4–8	7–20	14 tonnes	Cambio poles	(i)
4–8	12–30	12 tonnes	Butts	(ii)
4–8	12–30	14 tonnes	Butts	(ii)
1	7–30	12 tonnes	Weyroc	(iii)
1	7–30	14 tonnes	Weyroc	(iii)
2	7–30	12 tonnes	Weyroc	(iii)
2	7–30	14 tonnes	Weyroc	(iii)

b. Crane-equipped lorry loading other vehicles and produce categories:

Product specification		Gross vehicle weight	Product type	For terminal times see paragraph 6A, table number
Length (m)	Diameter UB (cm)			
1	9–25	32 tonnes	Bowater	(iv)
1	7–20	32 tonnes	PIM	(iv)
2	7–20	32 tonnes	PIM	(iv)

c. Produce stacked neatly on driver's side of the ride (or rack) in stacks of at least one full grapple load.

d. The average number and distance of moves when loading as indicated by the mean and range figures given in the tables in paragraph 6.

N.B. The Output Guide will not apply when these figures differ.

e. All forest roads, rides and racks travelled on to be free from obstruction and with no gradients of more than 10% (6°) over 100 m long.

f. Load weights not to exceed the legal limit but to be within 10% of that figure.

2 Job Specification

The Output Guide applies to the following work:

a. Loading of any lorry should be carried out in accordance with the 'Code of Practice for the Safety of Loads on Vehicles' published by the Department of the Environment.

b. *Self-loading and -unloading only*
Driving to the loading point by the shortest practicable route. Setting up bolster pins and preparing the Hiab crane. Loading the produce, moving between stacks of produce (including raising and lowering the crane and stabiliser legs). Securing the load and Hiab. Driving to the unloading point, weighing the load and unloading at the required place. Stowing all load securing equipment and the Hiab. Removing loose bark etc from the bed of the lorry.

c. *Loading other vehicles*
Driving to the loading point, guiding the customer's lorry where necessary, manoeuvring into position, setting up Hiab and stabilisers. Loading, including moving between stacks, raising and lowering crane and stabilisers as necessary, and the parking of the crane and stabilisers on completion of loading.

3 Tools and equipment (including safety equipment)

a. Lorry fitted with 2 tonne Hiab crane with log grapple unless otherwise specified.

b. Bolster pins as necessary.

c. Securing chains, dwangs or belts as required.

d. Pulphook and/or timber hook.

e. Safety equipment as specified by the Safety Officer, viz:
Safety helmet, complying with BS 2826 or 5240.
Boots with safety toecap.
Industrial gloves.

4 Allowances

Included in the Output Guide:

a. For contingencies and work other than that included in paragraph 2 above, eg daily maintenance, refuelling, washing, receiving instructions, repairs of up to 15 minutes duration, 5% of the time spent loading, travelling and unloading.

b. For personal needs and rest 16% of the time spent loading and unloading and 17% of the time spent travelling.

5 Method of using the Output Guide

Paragraph 1 above specifies the range of product types and lorries to which this guide applies.

The total time for the job is divided into two portions:

(i) *Terminal time*

Terminal time includes all work other than the actual travelling between the point where a full load is achieved and the reporting point at the delivery site and the return journey to the loading site. See paragraph 6A tables (i)–(iv) for times.

(ii) *Travelling time*

Travelling time is the time spent travelling between the loading site and the delivery point and the return journey. See paragraph 6B for times. Appendix II gives details of journey distances.

The appropriate terminal and travelling times are then added together to give the total time for loading, carriage and unloading.

Example

14 tonne Forestry Commission lorry
Product specification 4 m–8 m length × 12–30 cm diameter
Travelling distance for round trip   = 13 miles on public road
Travelling distance for round trip   = 3 miles on forest road
Conditions and job specification as shown in paragraphs 1 and 2 above.

Terminal time	= 67·0 SMs per load (from paragraph 6A table (ii))
Travelling time	= 2·5 × 13
Public road	= 32·5 SMs per load (from paragraph 6B (ii))
Travelling time	= 8·2 × 3
Forest road	= 24·6 SMs per load (from paragraph 6B (iii))
Total time per load	= 67·0 + 32·5 + 24·6 SMs
	= 124·1 SMs.

6A **Terminal time in standard minutes per load for lorries loaded and unloaded by their own 2 tonne Hiab crane**

(i) *Product type—4 m–8 m long × 7 cm–20 cm diameter UB*

	Loading site	Forest road/rack
Nominal carrying capacity of vehicle	12 tonnes	14 tonnes
Mean number of moves between heaps	3·7	4·6
Range of number of moves	2–6	2–10
Mean distance moved between heaps (metres)	87	103
Mean weight loaded (tonnes)	11·14	10·71
Terminal time in standard minutes per load	68·8	64·3

(ii) *Product type—4 m–8 m long × 7 cm–30 cm diameter UB*

	Loading site	Forest road/rack
Nominal carrying capacity of vehicle	12 tonnes	14 tonnes
Mean number of moves between heaps	4·1	5·3
Range of number of moves	1–9	2–15
Mean distance moved between heaps (metres)	28	65
Mean weight loaded (tonnes)	10·82	11·46
Terminal time in standard minutes per load	61·3	67·0

(iii) *Product type—2 m long × 7 cm–30 cm diameter UB*

N.B. Details on 1 m × 7 cm–30 cm diameter UB will be published in due course.

	Loading site	Forest road/rack
Nominal carrying capacity of vehicle	12 tonnes	14 tonnes
Mean number of moves between heaps	8·7	8·0
Range of number of moves	4–10	3–14
Mean distance moved between heaps (metres)	143	166
Mean weight loaded (tonnes)	11·65	11·09

	Loading site	Forest road/rack
Terminal time in standard minutes per load when self unloaded	113·0	136·0
Terminal time in standard minutes per load when unloaded by Volvo front end loader equipped with swivel clamshell grapple	106·5	126·2

(iv)   Terminal times for loading customers' lorries will be published in due course.

6B   Travelling time in standard minutes per load for all lorries and product types

(i)   Travelling through Brandon = 3·9 standard minutes per mile

N.B.

Distance through Brandon on the Elvedon road = 1 mile
Distance through Brandon from FR 2 = 3 miles
Distance through Brandon on the Thetford road = 2·3 miles

(ii)   Travelling on all other public roads = 2·5 standard minutes per mile

(iii) Travelling on all forest roads = 8·2 standard minutes per mile

See Appendix II for details of journey distances.

7   Modifications
None.

Working method

The normal working method to which the Output Guide applies when loading lorries is as follows:

a. Manoeuvre lorry close to heap of material to be loaded.

b. Place bolster pins at the required spacing.

c. Stand either in the lorry cab with the upper half of the body through the hatch in the cab roof or on the platform at the side of the lorry depending on the crane control layout.

d. Lower stabiliser legs.

e. Load lorry until no further produce is in reach.

f. Raise stabiliser legs and park grapple.
(Repeat a to f above as necessary until full payload is reached.)

g. Inspect load for stability and remove any branches etc which might cause a hazard on the public road.

h. Secure the load with chains or belts.

i. Drive along forest road to public road.

j. Stop at forest road, public road, junction. Inspect load for security and tighten down if necessary.

k. Proceed to delivery point at a speed which at no time exceeds any speed limits that are in existence.

l. On arrival at delivery point weigh lorry and proceed to unloading site as directed.

m. Remove load securing devices, lower stabiliser legs and unload lorry.

n. Raise stabiliser legs and park grapple, remove all debris from lorry bed and secure chains or belts to prevent them from falling off.

N.B. There will be occasions when lorries are unloaded by independent means at the delivery point. In these cases, apart from the actual removal of the load, the lorry driver's responsibilities will remain as outlined above.

Each way journey distances on public roads

a. Brandon Conversion Depot to:

	Miles
FR 47	3·8
Cranwich Camp	5·9
Lynford Cross	3·2
Top Santon Downham Road	4·1
Didlington Junction	4·8

a. Brandon Conversion Depot to: *Miles*
   Swaffham Office   6·1
   Hillborough   8·7
   C Cley   12·0
   FR 117   11·8
   Swaffham   14·4
   FR 105   17·8
   W Tofts   4·4
   New Buildings   7·0
   Wretham Camp   10·9
   FR 89   12·8
   Gt Hockham   13·6
   FR 93   14·9
   Bridgham Office   11·3
   FR 68   12·0
   FR 70   13·6
   FR 29   6·2
   Roundabout   9·4
   Peddars Way   13·6
   FR 76   15·6
   Mundford   4·0
   Santon Downham Village via Brandon   3·1
   FR 9 Brandon-Thetford road   4·0
   Traffic lights (Thetford)   9·1
   Weyroc via traffic lights   9·9
   FR 2   3·5
   FR 303   9·1
   Elveden Crossroads   5·4
   Mayday FR 6   2·6
   FR 203   8·5
   FR 207   11·9
   Swaffham Market   14·4
b. Weyroc Factory to:
   Traffic lights   0·8
   Elveden Crossroads   3·0
   FR 203   6·1
   FR 207   9·5
   FR 303   9·7
   FR 2   10·7
   Brandon   7·8
   FR 6 via Elveden   5·8

Printed in England for Her Majesty's Stationery Office
by McCorquodale Printers Ltd., London
HM 0104   Dd. 586086   K28   12/78   McC.   51–8924